STRONG IMAGINATION

The sleep of reason produces monsters, plate 43 of 'Los caprichos', 1799 (etching and aquatint). Goya Y Lucientes, Francisco Jose de (1746–1828). Private collection. The Bridgeman Art Library.

STRONG IMAGINATION

Madness, Creativity and Human Nature

DANIEL NETTLE

OXFORD
UNIVERSITY PRESS

Great Clarendon Street, Oxford OX2 6DP

Oxford University Press is a department of the University of Oxford.
It furthers the University's objective of excellence in research, scholarship,
and education by publishing worldwide in

Oxford New York

Athens Auckland Bangkok Bogotá Buenos Aires Calcutta
Cape Town Chennai Dar es Salaam Delhi Florence Hong Kong
Istanbul Karachi Kuala Lumpur Madrid Melbourne Mexico City Mumbai
Nairobi Paris São Paulo Singapore Taipei Tokyo Toronto Warsaw

with associated companies in Berlin Ibadan

Oxford is a registered trade mark of Oxford University Press
in the UK and in certain other countries

Published in the United States
By Oxford University Press Inc., New York

© Daniel Nettle, 2001

The moral rights of the author have been asserted

Database right Oxford University Press (maker)

First published 2001

Reprinted as TSP paperback 2001

All rights reserved. No part of this publication may be reproduced,
stored in a retrieval system, or transmitted, in any form or by any means,
without the prior permission in writing of Oxford University Press,
or as expressly permitted by law, or under terms agreed with the appropriate
reprographics rights organization. Enquiries concerning reproduction
outside the scope of the above should be sent to the Rights Department,
Oxford University Press, at the address above

You must not circulate this book in any other binding or cover
and you must impose this same condition on any acquirer

British Library Cataloguing in Publication Data
Data available

Library of Congress Cataloguing in Publication Data
Data available

ISBN 0–19–850706–2 (Hbk)

ISBN 0–19–850876–X (TSP)

1 3 5 7 9 10 8 6 4 2

Typeset in Minion by J&L Composition Ltd, Filey, North Yorkshire
Printed in Great Britain on acid-free paper by Biddles Ltd, Guildford & King's Lynn

For my parents
(Larkin was wrong)

Acknowledgements

I am most grateful to the Warden and Fellows of Merton College, Oxford, who gave me the opportunity to write this book. Thanks are also due to Professor Gordon Claridge, Professor Guy Goodwin, Dr Geoffrey Miller, and several anonymous referees for their advice on the content. I have greatly benefited from their views and comments, which have much improved the book. Inevitably there are a few places where I have not been able to follow their counsels, and all errors and speculations are mine alone.

I could not have written this book if it had not been for the teaching received over the years from three of my greatest mentors. They are Professor Leslie Aiello, Professor Robin Dunbar, and Dr Richard Passingham. They taught me to ask big questions in plain language, and to try to see the important simplicity in the complex answers. I am also indebted to Michael Rodgers, my editor at Oxford University Press, for his quick and unstinting belief in the project, and to Linda Antoniw, for her fine copy-editing skills.

Oxford University Press and I are grateful to Faber and Faber for permission to quote from 'This be the Verse', by Philip Larkin (from Philip Larkin, *Collected Poems*) and 'Examination at the Womb Door', by Ted Hughes (from Ted Hughes, *Crow*). We also thank the copyright holders of various illustrations, as detailed in the figure captions, for giving permission to reproduce material here.

Since writing this book I have become much more wary about saying lightly that something or other has kept me sane. None the less, my friends have been a continuing source of support in what has been a fraught time. Amongst my colleagues at Merton, I must express particular gratitude to Dr Matthew Grimley, Dr Senia Pašeta, Dr Simon Pulleyn, and Professor Suzanne Romaine. And this book's birth owes at least as much to my other life, at StageGraft/ COTG, with Bob Booth, Jonathan Gunning, Ed Hassall, Marianne Jacques, John O'Connor, Dominic Oliver, Alex Reid, and the rest. Finally, this was a time when old friendships could be drawn on, reinvented, and renewed; love and special thanks to Deborah Clarke, Tom Dickins, Kerry Elkins, Erica Klempner, Rachel Rendall, and Matthew Warburton.

D.N.
London, June 2000

Contents

Introduction *1*
1 From disease to difference and back again *12*
2 From nature to nurture and back again *36*
3 This taint of blood *58*
4 The storm-tossed soul *90*
5 The sleep of reason produces monsters *113*
6 Such tricks hath strong imagination *137*
7 The lunatic, the lover, and the poet *161*
8 Civilization and its discontents *187*
9 Staying sane *201*

Epilogue *213*
Further reading *216*
References *220*
Index *231*

Introduction

A Midsummer Night's Dream is Shakespeare's quintessential comedy. It is a tale of several very different sets of people whose lives become bizarrely intertwined in the course of a hectic summer night. There is a gaggle of lovers in conflict, the court of a king, and a crew of ludicrous commoners, as well as a whole kingdom of spirits. Within all of these groups, the key characters' goals are, for different reasons, frustrated. As they seek to right their problems, their situations become unwittingly and ever more strangely interlocked, until finally, in the inevitable denouement of a marriage feast, order is restored. The audience goes home happy.

However, behind the rather well-made comic plot, there is also a strange drama of imagination. The long night's machinations have left several of the characters pondering episodes of mind which, were we to look at them with a psychiatrist's eye, we would have to recognize as symptoms of madness. There is Bottom, a stolid weaver, who believes he has been transformed into an ass, waited on by fairies, and made love to by a queen. There are the young cavaliers, Demetrius and Lysander, who have experienced bewildering and unmanageable alternations of love, aggression, hatred, and more love. All is certainly not well in their mental lives. Oberon, the king of the fairy world, the overseer of both the characters' and the audience's well-being, has to resolve this problem. He does this by implanting into the characters', and the audience's, minds the idea that the disturbing episodes are nothing but a dream. As long as the characters can distinguish dream events from real events, then they are not mad, and the happy resolution of the play, and their lives, is possible.

However, this solution is slightly uneasy. It looks like sleight of hand, and I for one am left pondering where exactly the line is between madness and extraordinary experience. Where, for example, do the torrid emotions associated with love—the elation, the sleepless fervour, the nightmares, and the sweeping melancholy—give way to the clinical conditions of mania and depression? And where does the imagination of the playwright—who can create a world of fairies and men with asses' heads, and people it so vividly as to take us there with him—give way to the delusions of the psychotic, who needs medical help? Shakespeare, here as elsewhere, is attentive to the question of the boundaries of madness. In one of the most memorable speeches of the play, he gives the mortal king, Theseus, one account of the issue. For Theseus:

> The lunatic, the lover, and the poet
> Are of imagination all compact.

That is, the delusions of the madman, the fervour of those under the grip of strong emotions and of the creations of the artist, spring from a common source. He goes on to identify the common property in these cases as one of *imagination*; the ability of the mind to range beyond the sensory given into a vivid and fantastic realm;

> Lovers and madmen have such seething brains,
> Such shaping fantasies, that apprehend
> More than cool reason ever comprehends.

Theseus makes no claim either that all artists are mad, or that all mad people are merely misunderstood artists. The claim is, instead, that of a common imaginative capacity which, directed one way, can lead to madness, and, directed another, can lead to creativity:

> One sees more devils than vast hell can hold;
> That is the madman. The lover, all as frantic,
> Sees Helen's beauty in a brow of Egypt.
> The poet's eye, in a fine frenzy rolling,
> Doth glance from heaven to earth, from earth to heaven;

> And as imagination bodies forth
> The form of things unknown, the poet's pen
> Turns them to shapes, and gives to airy nothing
> A local habitation and a name.
> Such tricks hath strong imagination, . . .

What Shakespeare has posited, then, is a psychological trait, dubbed *strong imagination*. He has also made three claims about it. First, it is in an inherent aspect of human nature; second, it is especially highly developed in madness and in creativity; third, it is somehow associated with the business of love, or at least, sexual attraction. This book is an exploration of these three claims, and it will take us in directions even Shakespeare's remarkable imagination never could have presaged.

◆

In 1810, John Haslam, the director of London's ancient Bethlem mental hospital, published a book called *Illustrations of Madness*. This is thought to be the first book ever devoted to the detailed exploration of a single psychiatric case. The patient in question was one Mr Matthews, and the book was formed from both Haslam's observations and Matthews' own writings and sketches.

Matthews, in modern terms probably a schizophrenic, had been a permanent 'incurable' resident of the Bethlem for many years. However, apparently an articulate man, he had periodically persuaded panels of doctors that he was sane and should be discharged. Haslam's account aims partly to illustrate the extent of Matthews' insanity, as an argument for keeping him confined.

Matthews believed, for many years, that in an apartment near London Wall, there was a gang of seven villains who manipulated his thoughts and actions. This manipulation in general was called *assailment*, and it was done by means of a science called *pneumatic chemistry*. The chemicals involved were volatile magnetic fluids, which, when applied to the victim, caused a wide variety of dreadful symptoms. The pneumatic part referred to the means of projecting the fluids; the gang operated an *air-loom*, a large contraption of tubes and valves that could send a warp of fluid out to a distance of up to 1000 feet. The form of the loom was well specified; Matthews insisted that an account of it was to be found in the 1783 edition of *Chambers' Dictionary*, under 'Loom', and that its figure was to be seen on one the plates relating to 'Pneumatics'. In any case, Matthews provided a detailed sketch of the loom (Fig. 1), including a key to all the moving parts. He also had a sophisticated theory of how it worked; the victim must first be hand-impregnated with a priming dose of fluid, whereafter he would be susceptible to assailments from a distance, which worked principally on the central nervous system.

Figure 1 The air-loom, as drawn by Mr Matthews (by permission of the Bodleian library, University of Oxford, Douce HH130).

The assailments were of several types, determined by the chemicals involved and the settings of the loom, and Matthews achieved a systematic classification of them. *Fluid-locking* constricted the fibres at the root of the tongue, impeding speech; *cutting soul from sense* was a chemical blocking of the pathways from heart to brain, leading to a dissociation of emotion and intellect; *kiteing* lifted an idea into the brain of the assailed person and made it hang there, as boys hang kites upon the air, unbidden and distracting, to the exclusion of all other thoughts; *lengthening the brain* produced a distortion of whatever normal ideas the subject might be entertaining, so that they became grotesque, like the reflections in a hall-of-mirrors. Finally, amongst the several dozen possible assailments, *lobster-cracking* and *bomb-bursting* were murderous assaults on the whole nervous system, both likely to prove fatal. All of Matthews' psychological symptoms, and all the things he feared, were assignable to one or other of the gang's tactics to get him.

The gang consisted of four men and three women. Their descriptions read like the dramatis personae of a play. Bill, or the King, the chief and main operator of the machine, is unrelentingly villainous and has never been observed to smile. Jack the Schoolmaster, so-called because he records the gang's actions in shorthand, always seems to be shoving his wig back with his forefinger, and is often merrier than the King, saying 'I'm here to see fair play'. Sir Archy wears outmoded breeches and affects a provincial accent, although that is clearly not his true speech. Whenever Matthews challenges him, he replies '*Yho* are mist*e*aken'. He is low-minded and obscene, despite his affectations. Finally, there is the Middle Man. He is an engineer and manufacturer of air-looms. He is skilful in operation of the machine, and clearly takes delight in his cruel competence. He has a twang of the hawk about him.

Of the women, Augusta is small-breasted and sharp-featured. She corresponds with other gangs in the West End of town. At first she seems friendly and cajoling, but when she finds that she cannot influence or convince, she becomes exceedingly spiteful and malignant. Her temper has been getting worse and worse. In contrast to Augusta, Charlotte, a ruddy brunette, is kept at home by the gang, naked and poorly fed, and sometimes feels herself to be a prisoner. She is a steady, persevering sort of person, who is convinced of the impropriety of her conduct, but cannot help herself. Finally, there is the mysterious Glove Woman. She is frequently away, but when she is present, she is remarkable for her skill in managing the machine. The others, and particularly Sir Archy, are constantly bantering and plucking at her, like a number of rooks at a strange jackdaw. She has never been known to speak.

This gang is just one of a number dotted around London, who assail ministers and diplomats in pursuit of their nefarious purposes. Their agenda, on the whole, seems to be to surrender the secrets of Britain to the French, to republicanize Great Britain and Ireland, and especially to disorganize the British Navy. Matthews has become a particular target because he has gained knowledge of their plans, and threatens to uncover them to the authorities. However, by combinations of kiteing Matthews, lengthening his brain, and fluid-locking him, they can make him appear insane. Thus he will be locked away where no-one will listen to his warnings of the threat to the state. Only Matthews' great strength and intelligence, as he sees it, has allowed him to retain his insight into what is really going on.

Matthews was clearly mad. Needless to say, there was no such gang, nor any such science as pneumatic chemistry. However, Matthews' case does illustrate some important facets of madness which are central to the themes of this book. When a machine breaks down, one expects it to grind to a halt, and produce nothing, or nothing coherent. Similarly, one might imagine that the human mind, malfunctioning massively, would produce no mental life, or perhaps no coherent mental life, just a jumble of incoherent, non-specific, transitory parts of thoughts. This is what happens in the dementias, such as

Alzheimer's disease. In these conditions, the brain is 'smashed', and one by one, the mental faculties fall apart, break down, until the patient dies. In Matthews' case, though, this was not the pattern. Matthews was still quite healthy and alert, and not degenerating. There was no evidence that his memory or intelligence were impaired. His mental life was extremely rich and deeply imaginative. His beliefs about the world were evolving, but consistent, over a period of at least 13 years, and it is evident from his descriptions that he could recall and reflect on his experiences, including interpreting the interactions between people, with uncanny lucidity. His world was highly structured and coherent, and within its confines all phenomena were given reasonable explanations. It was, however, a world of utter madness.

Matthews' account could form the basis of a remarkable work of imaginative fiction. There is a plot. There are characters with clear purposes, and individualities that exceed those purposes. There are conflicts between them, and changes in their relationships. Given this lifelike quality, it is perhaps not surprising that certain doctors, those with whom he did not share the full extent of his delusions, from time to time assumed him sane. This phenomenon troubled Dr Haslam, who protested:

> Madness being the opposite to reason and good senses, as light is to darkness, straight and crooked etc., it appears wonderful the two opposite opinions could be entertained on the subject . . . because a person cannot correctly be said to be *in* his senses and *out* of his senses at the same time.

It seems to me that Haslam was wrong to oppose madness to sanity as light is opposed to darkness, for Matthews' mad world was very like life; as recurrent, as vivid, as full of system, purpose, and relationship as the real thing. For this reason, his world is highly believable as a story. Only for Matthews there was no denouement; no Oberon to tell him to wake and realize that it was all a dream. This is the central feature that makes Matthews mad and a story-teller not so; Matthews had no idea where reality ended and his imaginary work began. Indeed, such features of reality as his confinement in Bethlem and the death of Prime Minister Pitt were worked into the fantasy, and explained within its extraordinary scheme. Matthews' madness was not the opposite of reality; it was an alternative account of it, with some similarities to it and some differences, just as a fiction is an alternative account of reality. So, although Matthews could not simultaneously be out of his right mind and in it, his being out of his mind looks like a *version* of sanity, a version that one might be convinced by, at least until one had found out some facts about London in the 1800s, about pneumatic chemistry, and so on.

Psychosis, the technical name for madness, is a tremendously variable phenomenon; no one else could have had Matthews' delusions. However, certain features often recur, such as the attributions of thoughts and voices to imaginary individuals, and the invention of plots against oneself for which the most banal thing becomes evidence. We can thus take Matthews as an illustration of what an extreme psychosis can be like from the inside. This granted, it looks like Shakespeare was right to associate madness with strong *imagination*, and right to see the similarities between madness and the processes involved in imaginative creation more generally. Shakespeare's claim seems, on the basis of Matthews' experience, to be justified at the level of how madness *feels*. The rest of this book, however, will be concerned mainly with the relationship between madness, creativity, and the mind at the scientific level, the level of how it *works*.

◆

The existence of people like Matthews is a terrible puzzle. What causes some people to enter this deeply strange state? Obviously, the mind is going wrong, but as we have seen, that is not enough of an answer in itself, for two reasons. First, the mind is going wrong in a very specific, coherent, and imaginative way. Secondly, the things that happen in psychiatric malfunction are related to the things that happen in normal life. Delusions, like those of Mr Matthews, are related, at some level, to dreams, jokes, stories, and fantasies, which are common in the healthy mind and, indeed, highly developed in the finest minds. Mania and depression are related, in some way, to the healthy experiences of elation and grief. Those emotions are amongst the most exalted that a person can experience. It seems, then, that the symptoms of madness are related to the proper functioning of the human mind. They differ from health in that they are terribly exaggerated, unrelenting caricatures of their normal counterparts. It is as if the madman is locked in a hall of distorting fairground mirrors, which inflate grotesquely certain normal psychological impulses and abilities. Madness is not so much mental malfunction as a state of horrible hyperfunction of certain mental characteristics.

How can such a state exist? Surely the human mind evolved to give its user a faithful, useful, non-distorting model of reality and his or her place in it. How, then, can it be that there is a possible configuration of the mind where it leaves off from reality altogether and disappears into brilliantly complex, utterly useless, fantasy? And it is not as if madness is a very rare condition. It is known in all societies, and, in the societies well studied by scientists, affects around 2 per cent of the population in some form or other.

In this book, then, I will ask the question 'Why does madness exist?'. To attack the question, I must first investigate what the different forms of madness are, and what we know about the brain mechanisms underlying them. This is the work of Chapters 1–5, which form the first half of the book. In those chapters, too, I shall present clear evidence that genes are involved; the predisposition to madness is something one inherits along with eye colour and the ability to roll one's tongue. The discovery of genetic factors, though, does not answer my central question, except in the most limited of senses. At one level, madness exists because certain people have genes that predispose them to it. But this just raises the deeper question, 'Why do the genes that predispose people to madness exist?' One might think that natural selection would have eliminated them from the human gene pool, given that being out of one's mind can hardly be advantageous. Or could there be some hidden advantage?

Here I shall take Shakespeare's hypothesis seriously. He had Theseus propose a common psychological basis for the madness of the lunatic and the creativity of the poet. Now if the genes that predispose people to madness can also cause positive attributes such as enhanced creativity, then there would be a force keeping them in the gene pool. Madness would persist in our species because, although it is disadvantageous in itself, it is closely linked to a trait—creativity—which is highly advantageous.

This is the answer I shall be putting forward to the question of why madness exists. Shakespeare had no more than an anecdotal basis for proposing it, but it turns out, astonishingly, that the modern scientific evidence, which has only been amassed in the past 30 years, is squarely behind him. I present this evidence, and explore its implications for mental health, culture, and our view of ourselves as a species, in Chapters 6–9, which form the second half of this book.

◆

Shakespeare wrote *A Midsummer Night's Dream* in the last decade of the sixteenth century, although in fact the idea of a link between madness and creativity can be traced back to the ancient Greeks, if not further. It is an idea that has persisted powerfully in literature. Dryden, for example, wrote that, 'Great wits are sure to madness near allied/And thin partitions do their bounds divide'; Byron, more bluntly, 'We of the craft are all crazy'. It is also an idea that was much discussed in the early phases of the sciences of mind, which we now call psychology and psychiatry. Cesare Lombroso, Professor of Anthropology at Turin in the nineteenth century, argued in his book *Man of Genius* that lunacy and genius were flip sides of the same coin. Similarly, Benjamin Rush, who in 1812 wrote the first psychiatric treatise to appear in the United States, saw a commonality between the elevation of the imagination in madness and in 'talents for eloquence, music and painting, and uncommon ingenuity in several of the mechanical arts'.

One thing Lombroso, Rush, and literary authors lacked was systematic scientific evidence for a connection between cultural creativity and madness, and such evidence did not prove quickly forthcoming. Their theories were long left to languish, as psychology and psychiatry developed into more objective, rigorous, and professionalized disciplines during the twentieth century. This is perhaps not surprising; neither psychiatrists, struggling to cope with the very unromantic practical realities of dealing with mental illness, nor psychologists, keen to put their science onto a firm experimental footing, had, initially at least, much time for such airy notions. It is thus only more recently that scientific understanding of the link has developed, but the results show clearly that there was more than a grain of truth in the early speculations.

The second thing that was lacking until recently, and which prevented much progress being made in understanding the phenomena of strong imagination, was any real insight into how the human brain, the seat of such things, actually works. This situation is changing with astounding rapidity. In the second half of the twentieth century, we went from knowing little more than the gross anatomy of the brain, to beginning to understand how it is connected up, how its constituent cells transact their business, how to change moods and behaviours by administering drugs to it, and now even how to watch it in action through brain scanning. These advances, which are still under way, represent one of the greatest and most fruitful blossomings of knowledge in the history of science. They mean that the story of the extraordinary human mind—in brilliance and in sickness—is beginning to be legible, and the story is a remarkable one.

CHAPTER 1

From disease to difference and back again

> Who in the rainbow can draw the line where the violet tint ends and the orange tint begins? Distinctly we see the difference of the colors, but where exactly does the one first blendingly enter into the other?
> So with sanity and insanity.
>
> Herman Melville, *Billy Budd, Sailor*

This book is about mental disorders. In particular, I will be concerned with those severe mental disorders that come under the label 'psychosis'. *Psychosis* is the modern psychiatric term for what in earlier eras was called madness, insanity, or, implying a link to the phases of the moon, lunacy. It is that state of mind where a person's feelings or beliefs about himself, his fellows, and the world in general are completely disrupted, making him unable to function in whatever social role—husband, parent, friend, employee—he might expect to enjoy. It is the state where the sufferer passes beyond the bounds of reality, intelligibility, and rationality as defined by the bulk of society. The psychotic is a stranger among his own people.

Not all of the complaints that psychiatrists and psychologists deal with are psychotic complaints. A woman who is habitually depressed, anxious, aggressive, or unable to hold down a relationship, is not mad, although her problems may be real and serious. In earlier psychiatric terminology, a distinction was made between *neurosis*, which included these milder problems of living, and psychosis itself, true madness. In the latter condition, the break with normality and with the well person is complete, whereas, in neurosis, a basically whole person suffers difficulty or distress in some relatively self-contained area of mental life. The problems involved in drawing this line sharply, not least because people's expressed beliefs vary from situation to situation, are obvious, and will concern me greatly as this chapter progresses. Contemporary psychiatric schemes no longer include the category of neurosis. This is not the place to delve into the good reasons why the classification of mental disorders is subject to continual revision, for, in any event, the distinction between psychosis and the milder conditions is still maintained, the former still being called psychosis, and the latter being variously termed personality disorders and minor affective disorders.

Within psychosis, a traditional distinction was made between *organic* and *functional* psychoses. An organic psychosis has a clear physical cause. A previously healthy man's brain is penetrated by a fencing foil, or destroyed by disease, and he becomes demented. Material cause and behavioural effect are easily connected, and the psychosis is thus termed organic. In the functional psychoses, on the other hand, the madness appears without any detectable injury to the brain. In such cases one seems to be forced into one of two conclusions. First, one could conclude that the causes are psychological, not neurological. That is to say, they are to do with the functioning of the mind, not the brain, and one should look for explanations in the psychic domain, not the physical one. Perhaps the person is put under intolerable strain at work, or locked in impossible relationships at home. Secondly, one could assume that there is a subtle brain change, the result not of massive injury from without, but perhaps of insidious changes in the brain's self-regulation, and one could look for the biological traces of that. As I will argue in this book, both of these alternatives are valid.

More importantly, the notion that there is some kind of dichotomy between them is quite misleading. To call the causes of madness psychological not neurological is to imply that there are two distinct kinds of causes of human behaviour, psychic causes and physical causes. But this is not so. The mind is not a different thing from the brain. The mind is simply what the brain does. That is to say, psychic events are no more than a way of looking at brain processes, the way that we look at those processes from the standpoint of everyday human experience. There is no subjective sensation or thought which is not also a brain process. Thus, psychosis is not caused by brain changes as such; rather it is those brain changes, and the psychotic symptoms are what those brain changes do from the person's point of view. It follows that any psychosis is a physical process. Thus it can be caused by any other physical process or event, be it a fencing foil, a hallucinogenic drug, or a prior brain state. This earlier brain state might, in turn, be caused by something in the social environment, since the activity of the brain is constantly changing in response to input from the senses and the body.

In the past two decades, it has become increasingly clear that there are subtle brain changes associated with what had been called the functional psychoses. Many investigators, using several techniques, have found such differences. Unfortunately, though, they have not consistently found the same ones. The changes have been detected at several levels. There seem to be some differences in the anatomical structure of the brain, detectable by a scanning technique called computerized tomography. By way of example, several studies have found both schizophrenic and manic-depressive patients to have an enlargement of the hollow ventricles that lie deep in the brain. There are also reports of differences in patterns of brain functioning. These are detectable using another type of brain scanning, called positron emission tomography (PET), which instead of giving an image of the anatomical structure of the brain, gives a picture of its activity at the time the subject is scanned. PET studies have revealed uncharacteristic patterns of activation in the scans of both manic-depressives and schizophrenics. Several studies have found depressives to have increased activity, compared to normal subjects, in a brain structure called the amygdala, which is known to be associated with emotion. This is true both in those who are currently depressed, and in those who have a history of depression but are currently in remission. The abnormal activity of the amygdala is thus a reflection of the underlying depressive temperament, not the current state. In those whose symptoms are currently bad, there is also increased activity in some parts of the left frontal lobe, and decreased activity in other parts. These changes seem to be the brain manifestation of their current flare-up.

Finally, the levels of certain chemicals internal to the brain, called neurotransmitters, appear to be abnormal in psychosis. The levels of these chemicals are difficult to measure directly, for obvious reasons. Their functioning can, however, be detected indirectly through their consequences for the chemistry of the blood, urine, and cerebrospinal fluid. Many, although not all, investigations of psychotic patients have found the operating level of neurotransmitters to be abnormal. Furthermore, the drugs that are now the treatment of choice for psychosis work by augmenting or stabilizing the levels of specific neurotransmitters, and their great effectiveness strongly implicates these substances. In view of scientists' new understanding of such subtle brain changes, the authors of one review recently argued that the distinction between the functional and the organic psychoses has become less meaningful in recent decades.

It is true that there are subtle brain changes in the functional psychoses, but this does not abolish the functional/organic distinction entirely. This is because, as I have already said, the brain changes detected have not been shown to be the causes of the psychosis. Rather, they *are* the psychosis, considered at the physical level. This is not quite the same as in the classic organic psychoses, where the organic trauma (the fencing foil, for example) is clearly prior to the psychosis and distinct from it (and therefore can stand to it as cause stands to effect). So, in short, the discovery of the brain mechanisms of the functional psychoses, about which I have much to say in later chapters, does not explain their origins. They could still be caused by stressful relationships, or genes, or diet, or any of the other numerous inputs that determine a person's current brain state. Questions of causality thus remain open for functional psychoses in a way they do not for organic ones.

I will retain a working distinction between functional and organic psychoses, then, but the reader should be careful not to take it too literally. To do so would threaten a slide down the slippery slope towards some kind of dualism. Dualism is the philosophical position that holds that there is a distinction in kind between mental things and material things; or, if you like, between functional madness and organic madness, psychic phenomena and physical ones. It assumes that different laws of causality apply to each sphere. Now, the seduction of dualism is an unwitting one. Few, if any, psychiatrists have ever believed that a naive dualism of psyche and brain is literally true. The functional/organic distinction is just a practical way of grouping their cases, not a theory about the world. However, the distinction between mental and physical talk is an easy and natural one to slip into, and sometimes having done so, people get carried away with it into strange and dualistic arguments about mental illness, as we shall see in a later chapter.

◆

The functional psychoses are generally divided into two groups: schizophrenia and affective psychosis. The distinction between the two types, and the modern classification of mental illness more generally, has as its most obvious parent the German psychiatrist Emil Kraepelin. Kraepelin's work is worth dwelling on, both because of its seminal nature and because it allows a useful glimpse of some important landmarks in the history of psychology.

Kraepelin was probably the most influential psychiatrist of the nineteenth century, although his name is not widely known outside psychiatry. This is in sharp contrast to his psychoanalyst contemporary, Sigmund Freud, who is universally recognized. The contrast between the two men is instructive, and illustrates the way the study of the mind has tended to split along lines that have a suspicious whiff of dualism about them. Freud, the interpreter of dreams and of unconscious conflicts, is usually associated with the psychic or humanistic side of the dualism. His work dissects the mental life of the individual patient, unfolding its hidden meanings and tensions through individual case histories that, by Freud's own admission, read more like short stories than works of science. Kraepelin's objectives were quite different, and he is associated with the physical or biomedical side of the dualism, the view that mental problems are organic diseases like any others. Kraepelin was determined to achieve some kind of systematic classification of the different forms of madness, in order that patients could be assigned to a diagnostic category and the correct treatment applied. He studied thousands of patients, and he was clearly not much interested in them as individuals, or in the content of their mental lives. They were better viewed as clusters of symptoms, which could be assigned to one category of disease or another. The main task for the psychiatrist was diagnosis and medical treatment; for psychiatric research, to classify, to seek the physiological mechanisms of disease, and to look for causes. On the question of causes, Kraepelin always suspected heredity, and it was he who initiated the modern study of genetic factors in psychosis.

These observations are not criticisms of either the Freudian or the Kraepelinian tradition. They are also an oversimplified rendering of the similarities and differences between them. However, most of the positions seen in psychiatry in the past 100 years can be situated somewhere along the road between an idealized Kraepelinian biomedicine and an idealized Freudian humanism, and so this is a useful distinction to make. The tendency to assume that people with manic-depressive disorder, obsessive-compulsive disorder, or schizophrenia each have a distinct brain abnormality which must be detected through characteristic symptoms and then treated with the appropriate drugs is often termed 'neo-Kraepelinian'. By contrast, the various alternative schools based on the therapeutic exploration of the individual's life experience, which care little for diagnostic categories or for drugs, are more obviously in the general lineage of Freud, although not necessarily Freudian in any narrow sense. This is, as I have said, simplistic; most contemporary practitioners fall into a middle category, a category that might include both Freud and Kraepelin were they still alive.

None the less, it is of great use to the present story to keep in mind the two great traditions in mental medicine, the humanistic and the biological, locked as they are in an eternal double helix. The topic of psychosis and creativity is, oddly enough, one of the few areas where they fuse and run into each other, for investigators in this area, as we shall see, have had to combine concerns about brains and statistics with a more qualitative interest in what creativity and psychosis feel like from the inside.

Kraepelin's achievement was to bring together a large and indeterminate number of syndromes under two broad umbrellas. He called the first one 'manic-depressive insanity' (now generally known as affective psychosis), and the second one 'dementia praecox' (now called schizophrenia). By 'dementia praecox', he meant a kind of dementia which came on early in life, in contrast to the dementias of the geriatric, which are mainly organic psychoses related to the ageing of the brain. The exact borders of the Kraepelin's two concepts have changed over the 100 years of their existence, and are still disputed, but their broad usefulness is still accepted in most quarters. The only change that is of importance here was brought about by Eugen Bleuler in 1911; he somewhat broadened the definition of dementia praecox, and renamed it with its more familiar designation, schizophrenia. He chose this term, from the Greek *schizo*, meaning split, since he believed that the core of the syndrome was a splitting, be it between different parts of the mind, between the reason and the emotions, or between the self and the real world.

Schizophrenia is madness in its pure form. There is no single symptom, but rather a constellation of related abnormalities, which can be divided into positive and negative categories. The positive abnormalities are disruptions of thought processes such as delusions or hallucinations. The vivid fantasies of Mr Matthews, described in the Introduction, mark him out as a probable schizophrenic. Both the form and the content of schizophrenic fantasies are enormously variable, but the classic type consists of 'hearing voices' in the head, which are thought to belong to other people. On the negative side, schizophrenics often show a flattened emotional response to those around them, seen as coldness or indifference. They can also have deficits of motivation and of reasoning, and other intellectual abnormalities which I will discuss in Chapter 5. The negative symptoms are important, but a positive disturbance such as hearing voices is required for schizophrenia to be diagnosed.

The positive symptoms of schizophrenia classically come on in young adulthood, although there is a lot of variation in this, and the average age of onset is a little earlier for men than for women. Some patients have one psychotic episode and return to relatively normal life without relapsing. In other cases, the disease continues chronically throughout life, with intermittent episodes and remissions. Even in these days of antipsychotic drugs, the prognosis for many schizophrenics is fairly poor. Around one-quarter return to completely normal functioning; about the same number again make an intermittent or partial return to work and social independence. Many of the rest fall into chronic illness. The lifetime risk of schizophrenia is generally estimated at around 1 per cent.

In contrast to schizophrenia, where the disturbance starts from thought, the affective psychoses are, initially at least, imbalances of emotion. This earns them their name, from the Latin *affectus*, or feeling. I refer to the affective psychoses in the plural because they have several components. Very severe depression is an affective psychosis, as is the condition commonly known as manic-depression, where the sufferer oscillates between severe depressions and periods of mania. The former is called unipolar affective psychosis, as the emotional disturbance is all in one direction, whereas the latter is called, for obvious reasons, bipolar affective psychosis.

The classical picture of the depressed state is familiar to most of us. The sufferer has excessive and lasting feelings of sadness, gloom, or worthlessness. His energy is sapped, his posture is rounded, and even everyday tasks present impossible setbacks. He is lethargic. His outlook is negative, his application desultory, and his judgement about what he can achieve pessimistic.

This classical picture is common, but it is not the only way in which depression can manifest itself. As well as versions of depression in which sadness and resignation are the predominant themes, there are versions where anxiety, anger, or fear predominate. There is even some evidence linking such diverse phenomena as alcoholism, eating disorders, impulsive criminality, and violence to the same brain mechanisms as classical depression. The common theme in all these states is a predominance of negative feelings, and a diminution or absence of positive ones. A deficit of positive feelings is called *anhedonia*, and it is a defining feature of depression. The difference between the versions of depression resides chiefly in the way the sufferer responds to the negative feelings, whether accepting with resignation (classical depression), lashing out (impulsive violence), seeking solace (alcoholism), or taking it out on the self (eating disorders).

As well as feelings, really severe depression can permeate external beliefs. This is when it becomes an affective psychosis, as opposed to a non-psychotic affective disorder. The sufferer is overcome with feelings of guilt or inadequacy, which may spread out into deluded beliefs. These range from the mild, such as the belief that friends have secretly turned against him or been soured by her, to the bizarre, such as the idea that he has personally caused the Second World War or a major natural disaster.

The flip-side of depression is mania. Mania is an enduring state of great elation. The subject feels high on life, and able to take on anything. He makes grand plans for ambitious future projects, sometimes spending vast sums of money, entering into risky ventures or novel liaisons without constraint. He has no need for sleep or the conservative platitudes of his fellows. His new ideas about the world are sweeping, often grandiose, sometimes flowing into each other, sometimes leap-frogging in a delicious flight which is logical to him but baffling to his companions. Mania, too, becomes psychotic, at the moment where the division between mere ideas and actual beliefs becomes blurred.

Depression is something we can all relate to, and in its non-psychotic forms it is extremely common. Indeed, there is an obvious prototype in the universal emotion of unhappiness. However, the disorder is much more than this. It is an unhappiness that carries on in the absence of immediate reasons for unhappiness, and indeed persists despite participation in previously pleasurable experiences. There is clearly a gradation of severity of such mood alterations. The extreme cases are difficult to envisage for those who have never known or read about them, for the bleak mood becomes a massive, all-pervading torrent, flowing into every cranny of the subject's life at every instant. It overwhelms all mundane activities, all interactions with others, and, eventually, all the subject's beliefs about the world. It is only these severe cases that count as psychoses. It should be noted here that the term 'psychotic' is used in different ways by different authors. Sometimes, a depression is called psychotic to indicate its great seriousness. Elsewhere, its use indicates specifically that the subject suffers from delusions or hallucinations, which occur in a minority of serious depressives. This second usage is thus a subset of the first; the difference between them is fairly unimportant for present purposes.

Estimates of the prevalence of affective disorder vary widely according to the population sampled and the methods and the diagnoses used, but the general finding is a lifetime prevalence rate of 8 per cent or more, with the classical depressive syndrome occurring about twice as often in women as in men. Some, but not most, of the most serious depressives are bipolar, which is to say they experience mania as well. There are reasons for thinking that the bipolar form and the unipolar form of the disorder are related, and that the bipolar form is the most severe manifestation of a morbid continuum that runs from minor depression and discontent, through major depression, to manic-depression. Full-blown bipolar affective psychosis affects about 1 per cent of the population, a similar risk as for schizophrenia, and, in contrast to unipolar depression, there is no gender imbalance. There are also some rare cases of unipolar manic psychosis.

In the classic bipolar illness, episodes of depression lasting perhaps several months alternate with periods of normality and episodes of mania. However, there are also similarities between the two states, both in terms of brain functioning and of experience, and the sufferer can often feel the heightened agitation of the moods in bewildering succession, or almost at the same time. In these mixed states, acts of great recklessness are common.

The long-term prognosis in affective psychosis is somewhat better than that for schizophrenia. This is partly due to the efficacy of lithium and modern antidepressants, but partly predates the use of these drugs, and stems from the fact that the disease seems, on average, to be less all-pervading and more remitting. In early psychiatric texts, affective psychosis was called '*la folie circulaire*', to emphasize the cyclical nature of its appearance. The sufferer often functions at normal level, or better, between the cycles, and the well periods can be long and frequent. None the less, affective psychosis is a deadly tyrant, and its chief executioner is, cruelly, the sufferer himself. Around one-fifth of all bipolar patients end their lives by suicide, and at least two-thirds of all suicides recorded are committed by people previously diagnosed with an affective illness. (However, suicide is not uniquely diagnostic of affective disorder, at least 10 per cent of schizophrenics also end their lives this way.)

I have taken it for granted so far that schizophrenia and affective disorder are quite distinct, one beginning from thought and the other from emotion. However, it is not quite obvious that this is true. The emotional coldness of the schizophrenic may not mean lack of emotionality, but rather the experience of inappropriate emotions, or conflicting emotions at the same time. This is strangely reminiscent of the bipolar patient in a mixed state. On the other hand, delusions and hallucinations are not restricted to schizophrenics. In the extremes of mania and depression, frankly deluded beliefs are common, as when the sufferer believes that he personally caused the Second World War. Such a belief is not too far from the fantasy world of Mr Matthews.

There are areas of overlap, then, between the two functional psychoses, and drawing a diagnostic line between them has not proved easy. Most schemes recognize a middle category, known as schizoaffective disorder, which displays aspects of both. There is also considerable heterogeneity within each of the two categories, especially schizophrenia. It may be better to think of schizophrenia and affective psychosis as clusters of points on a map of psychotic functioning, clusters that may overlap. This is not to prejudge the question of whether the two clusters stem from a single cause, to which I will return in a later chapter.

◆

The classic psychiatric approach to schizophrenia and affective disorder is to consider them as diseases like any other. This view has certain implications, as has often been pointed out. For one thing, it casts the psychotic experience as negative and undesirable. This is not perhaps very surprising, as a condition attended with suffering and with such disastrous personal consequences can hardly be welcomed. However, it is worth remembering that many manics and schizophrenics do not consider themselves ill, and, as we shall see, there are movements that have stressed the healing, even the visionary, aspects of the psychotic experience.

The more serious consequence of the disease view, for which it has been much criticized, is that it invites us to think of the psychosis as something other than the person undergoing it. A person suffering from cholera is hosting a bacterium, which operates on his body and causes certain signs and symptoms of malfunction. The physician, where possible, removes the bacterium, and the symptoms disappear. Importantly, the operations of the parasite are not those of the patient's body, but are alien to it, completely discontinuous with its normal functioning. He, as an individual, is no more personally implicated in the effects of his cholera than he is implicated in the effects of the bacteria that sour the milk in his refrigerator. And the implication of the disease view seems to be that the same is true for psychosis.

This implication squares with some experiences of mental disorder. Many psychotics feel that their illness is a thing alien to them, and resent its intrusion into their lives. Major depression, for example, is such a savage, inexplicable experience that many sufferers have felt it to be quite distinct from their prior mental life. Elizabeth Wurtzel, for example, writes in her book *Prozac Nation*:

> That's the thing I want to make clear about depression: it's got nothing at all to do with life. In the course of life there is sad-

ness and pain and sorrow, all of which, in their right time and season, are normal—unpleasant, but normal. Depression is in an altogether different zone.

Similarly, we cannot read the fantasies of Mr Matthews without feeling that there is a radical discontinuity, a definite and distinct pathology between his mental world and the normal one. Or can we? And is this only a product of considering just the most extreme cases?

In sharp contrast to the disease perspective stand a wide variety of approaches that view mental disorders not as diseases but as simply the expression of personal *difference*. It is easy to see where the logic of such philosophies comes from. Let us take the example of psychotic depression once again. Sadness is a universal human experience, often an appropriate one, and a minor depression is simply a lasting sadness. For every minor depression, it is quite possible to conceive of a slightly more enduring, deeper, more all-pervasive version. And then there is a slightly more serious version of that one, and so, in turn, up to a full-blown depressive psychosis. Thus the most severe state of depression can be linked via a chain of intermediate positions to normal functioning. This is not to deny the severity of the symptoms of the psychotic condition. Rather, it is to acknowledge that human emotional states exist along a continuum. It is a common observation that people differ in the extremity of their emotional reactions to life events, and the susceptibility to depression may just relate to this continuum.

There are two ways in which I might attempt to refute the continuum argument, and maintain the view of major depression as a categorically distinct entity. First, I might distinguish between lowered mood which has a comprehensible cause, and lowered mood which appears to be a pathology. For example, a woman bereaved of her husband might be reasonably expected to enter a period of chronic sadness, pessimism, and lethargy. If she does so, disease does not seem an appropriate concept to invoke; rather it is a natural, if distressing, process. On the other hand, if a successful and newly married woman had the same symptoms without identifiable cause in her life, the language of illness seems more attractive. I might therefore keep the discontinuity concept, but only employ the category of disease where the lowered mood is *unjustified*.

The idea of justification cannot save the discontinuity argument, though. People differ enormously in the strength and symptoms of their emotional reactions to the same events. If the widow in my example continued to mourn deeply for 2 years, I might just say she had been strongly attached and was of reactive personality. After 5 years, or 10, or 15, I might suspect pathology rather than normal grief. But where is the line to be drawn? This problem is exacerbated by the fact that different cultures have very different norms relating to the appropriate display and timing of grief. From the perspective of a Jewish family, with their tradition of mourning for a year, then lifetime grief might seem to be pathology. But for traditional Greek and Portuguese women, this was just what was expected. To look at it another way, a bereavement is obviously sufficient justification for a profoundly lowered mood, but there are other cases that are less clear cut, such as a disappointment in a very brief love affair, or the loss of a pet. A massive reaction to these latter events might lead us to suspect an underlying depressive disease for which the events are just a rationalization after the fact. But once again, where is the line to be drawn? It is clear that between normal, justified mood fluctuations and major depressive illness, there is nothing but a continuum, and different individuals and different cultures occupy different places along the scale.

The other way in which the discontinuity view might be defended is by pointing out that in the full-blown psychoses, the individual has acquired beliefs which are, by the common consent of his culture, untrue. That is, he is deluded. Thus the depressive psychotic who believes he is personally responsible for all the suffering in the world is qualitatively different from the person who merely feels enduringly unhappy and worried. This argument is superficially attractive, for the two cases do seem to be quite different. However, I can, once again, easily envisage a variety of intermediate cases. This is because we *all* have many beliefs that go beyond the truths generally acknowledged by our cultures. When depressed I might believe myself to be an unattractive failure, and my work to fall far short of its potential value and quality. Now it is hard to judge whether these beliefs are true by the standards of my culture or not. Some people find me unattractive, and perhaps these are the people I have in mind in my belief. Similarly, some people like my work and some not, but no one has the same sense as me as to what it could have been if I had been as clever or as diligent as I would like. Finally, no one can say definitively whether it is a true statement that I am a failure or not, because they do not know the (perhaps unconscious) goals and standards that drove me on. In short, such beliefs are of annoyingly indeterminate truth value.

Now one might counter that the true psychotic has deluded beliefs about the objective world, whereas the cases I have discussed above are all beliefs about myself. This argument is not straightforward either. Statements such as 'The world is getting worse and worse' and 'There is no place for someone of my limited abilities in this business' are commonly uttered by depressives, yet they are about external reality, not the subject's self. And it is hard to say straightforwardly whether they reflect reality or not. This problem is hugely exacerbated by the cultural relativity of acceptable beliefs about the world. If I, as a British person, believe that a dead ancestor lives on in spirit and speaks to me through a soft egg, then I am held to be deluded. However, in certain central Nigerian cultures, such a belief is normal. Similarly, hearing voices is a psychiatric symptom in a doctor's surgery, but quite commonplace in a Pentecostal church. In fact, many bereaved persons experience hallucinations of their lost companion's voice or appearance. This probably stems from long habit, and is not sufficient to suspect disease. But how much of this, and for how long, can we accept before invoking pathology?

I could make the same point another way by invoking the differences between people. Some people are of characteristically gloomy outlook. They often feel that it is difficult and enervating to get the things they want, and they may be pessimistic about the possibility of changing what they judge to be an unsatisfying life. They thus risk little, concentrate on modest and immediate goals, often feel tired by life, and avoid too much change. Some people, by contrast, are sanguine risk-takers, always ready for new experiences or new challenges, and confident of their abilities. There is thus no unique baseline for human mood. More importantly for the consideration of bipolar disorder, there is no unique baseline for the amplitude of emotional fluctuations people experience in response to life events.

This insight has been formalized by the branch of psychology that deals with individual differences. Psychology measures personality in two general ways. First, there are 'soft' measures which are established from the subject's responses to questionnaires. The development and study of these questionnaire measures is called *psychometrics*. Secondly, there are more 'biological' measures which directly test the reactivity of the subject's nervous system to different kinds of stimuli. The measures used include the conductivity of the skin (the galvanic skin response), the pattern of brain electrical activity (the electroencephalogram, or EEG), and the threshold of response to various drugs and sensory stimuli. Encouragingly, studies have shown reliable correlations between subjects' reporting of themselves on the questionnaires and their actual reactivities on the biological tests.

Psychometric studies have indeed shown that in the population at large there are wide and continuous spectra of personality variation on traits such as emotional negativity, extraversion, and impulsivity. Extreme scores on some of these scales are associated with increased risk of psychiatric disorder, and, crucially, levels of the very same neurotransmitters that have been implicated in psychosis seem to be related to the position of an individual on the personality spectrum. This has been shown both by studies that monitor blood levels of relevant chemicals, and by studies that manipulate neurotransmitter levels experimentally, using drugs. Subjects who can be inferred (from their biochemical reactions) to have high activity of the neurotransmitter dopamine also appear as outgoing and seeking positive emotional experience. Meanwhile, deliberate depletion of the neurotransmitter serotonin makes subjects hostile or gloomy, and the administration of Prozac, which enhances the function of serotonin, makes normal subjects more positive and sociable.

The lesson of psychometrics is that the same symptoms, which might count as evidence for depression from a person of one personality disposition, might seem fairly normal from a person of another. It seems rather bizarre to argue that the same symptom could be disease when it happens to me but normality when it happens to my neighbour. And indeed, if I observe that my neighbour has a gloomy disposition, it is unclear that I have the right to call him ill just because his expectations are different from my own. Perhaps that is just how he is, or even how he wants to be. There is even some evidence that depressives are actually *more accurate* at judging their professional competence than the undepressed, who consistently overestimate their own abilities. To treat a person's outlook as an illness would seem unjustifiable, a kind of psychological imperialism; but the Prozac studies show that this drug affects nothing other than a selective manipulation of certain aspects of the human personality.

It seems, then, that although the extremes of health and disease are easy to identify and very obviously different from one another, there is a large grey area in between. The great strength of the difference perspective is that it acknowledges that the lines between the major mental illnesses and normal mental life are not sharp discontinuities in nature, that leap out at us. Rather, there is a spectrum of psychology onto which we, as a society, draw a murky line as best we can. What we call mental health, and what mental illness, is partly a decision rather than a discovery. At the psychotic ends of the spectrum, the distinction is obvious enough, but even psychosis is just that, the end of a spectrum.

The difference and disease perspectives clash heads most dramatically over the issue of schizophrenia. Schizophrenia, with its delusions and hallucinations, seems at first glance to be a clear example of a discontinuity with normal mental life. However, further research makes this assertion more problematic. For one thing, there are milder and more transient cases of schizophrenia-like symptoms; these are not considered full psychoses, and psychiatrists classify them as schizophrenia-like personality disorders. These intermediate cases have led some to talk of a schizophrenia 'spectrum' rather than a discrete category of disease. Furthermore, numerous studies have shown that occasional hallucinations are quite common in the normal population. In one study, 39 per cent of American college students reported having heard their thoughts spoken out loud, and 5 per cent had had conversations with the voices. Those subjects with a strong interest in music, art, poetry, and mathematics were particularly prone, and only 29 per cent could report no experience of a hallucinatory type at all. The prevalence of hallucinations was even higher in studies where the experimenter tried to bring hallucinations on, whether by hypnosis or by simple suggestion. What is more, these results were all from settings in which hallucinations are generally downgraded and stigmatized, as they are in the contemporary West. Most other cultures have a tradition of hallucination, which is often regarded as healthy and even blessed.

If hallucination is the rubicon of schizophrenia, then, schizophrenics must differ from normal individuals by degree, not in kind. Delusion could be argued to be the criterion instead, but it is not evident exactly what constitutes a delusion. Thousands of people routinely believe in astrology, the paranormal, tarot, and the spirit world, but otherwise function quite normally. Their beliefs may be shared within subcultures, but are often idiosyncratic and not supported by objective evidence. The normal precursor of schizophrenia is divergent or unorthodox thinking, and this is clearly a graded and relative concept. The belief that bread can turn into the body of a long-dead teacher when you eat it, or that there is somewhere an omnipresent deity able to read one's mind, is not, on the face of it, less obviously deluded than the ravings of Mr Matthews. It is just that the former beliefs happen to have worked their way to the top of the pile in Christian cultures. Thus it seems that society licenses some deluded beliefs, and does not permit others. From this perspective, those labelled as schizophrenics are simply those who refuse to share the widespread beliefs of their society; they are not diseased, just non-conformist.

This view was developed most famously by the maverick psychiatrist Thomas Szasz. For Szasz, there was no such thing as mental illness. According to him, an illness is an identifiable pathology of the body. This is not, he argues, what we are dealing with in schizophrenia. Rather, schizophrenia is simply deviance from the norms of belief and value that society demands. For him, then, 'schizophrenia' was not a medical problem, but a social and legal one, and schizophrenics were no different in principle from conscientious objectors or separatist zealots. Szasz was critical of organized psychiatry, which he felt to be a totalitarian institution for labelling and controlling those who chose not to live by the norms of their time, rather than a real branch of medicine. He and other writers critical of psychiatry made much of the fact that in the Soviet Union, those who disagreed with the prevailing ideology were diagnosed as schizophrenic and locked up for 'treatment'. Szasz also compared psychiatric diagnosis to the Inquisition, which controlled so-called heretics in the late Middle Ages; a totalitarian and purely political mechanism for controlling social diversity.

Szasz's arguments, and those like them, date from the 1960s and 1970s, when the enormous, grim asylums of 100 years ago were still a fresh memory, and when the prevailing intellectual climate was charged with progressive and deconstructive fervour. Szasz's target was those vast institutions that locked people up for decades without much consultation or negotiation. Such institutions have, by and large, disappeared; however, the problem of schizophrenia has not.

Szasz's argument might appear to be undercut by recent findings that schizophrenia is reliably associated with differences in brain functioning, as revealed by brain scanning, and can be treated with drugs that suppress specific brain chemicals, such as dopamine. Given that this is so, then surely schizophrenia is biologically real, not just a fiction invented by society to control dissidents.

This reasoning does not quite work. No biological differences have been found that have any diagnostic validity. That is to say, on average, schizophrenics have larger cerebral ventricles than non-sufferers, but there is a great deal of variability between individuals and no one could definitely say, by looking at a brain scan, whether the brain was schizophrenic or not. There is a continuum of ventricle size, with psychotics towards the top end. Similarly, although there are thought to be neurotransmitter imbalances in schizophrenia, there is also a spectrum of neurotransmitter function within the normal population, and this relates to the personality. Even if a level of neurotransmitter was found, beyond which psychosis was always the result, it would not prove that disease was the appropriate concept to invoke. Some people are taller than others. This is an objective biological difference, and those who are extremely tall or extremely short have certain practical problems of living. However, we do not thereby conclude that extreme height is a form of disease, and we certainly would not consider the involuntary hospitalization of the tall. And tranquillizers make convicts docile; but it does not follow that criminality is an illness caused by tranquillizer deficiency. The neurobiological evidence may, or may not, eventually reveal the basis of the difference between schizophrenics and other people. It cannot, however, answer the conceptual question of whether that difference should be classed as disease. Szasz's argument, then, is not so easily dismissed.

However, Szasz's position is no longer taken seriously in psychiatry, and when you look more closely at it, you can see why. It is wrong because it starts from a faulty conception of the meaning of 'disease'. It is true that the underlying disease process in schizophrenia is not known in the way that the underlying process is known for, say, cholera. This just means that schizophrenia is a *syndrome*—identifiable by a set of symptoms, rather than a cause. In this it does not differ from chronic fatigue or irritable bowel, and those are fairly obviously medical. Indeed, historically, most diseases were just syndromes for a long time, until someone discovered an underlying pathology. At this point, the classification of the diseases was often revolutionized, and this may yet happen with the psychoses.

It is also true that a continuum of functioning has to be recognized, with the functional psychoses as just the very extreme points and normality in the middle. However, this does not mean that the functional psychoses do not exist. That would be like saying that there is no difference between a snowflake and an avalanche because there is a continuum from a single snowflake to 100 000 tons of snow. Avalanches and snowflakes are importantly different in their human effects, and this is so with psychosis. Where the line should be drawn exactly is a sensitive matter for medical science, which will probably never be resolved once and for all. But this does not mean that the responsibility of drawing it should be shunned. There are though, as Szasz is right to stress, special problems with schizophrenia which do not apply in most cases of disease, which stem from the fact that most schizophrenics do not acknowledge themselves to be ill. Where their beliefs should be overridden, and where respected, is a terrible problem for any liberal society, and would take us off into much wider issues.

The main problem with Szasz's anti-disease stance is that it rests on a completely false notion of what disease is. For him, it is clear that a disease is always something like cholera, where a discrete physical cause external to the sufferer is responsible for a discontinuous abnormality in bodily function. But not all organic diseases are of this sort, as Gordon Claridge has pointed out in his book *Origins of Mental Illness*. There are also the *systemic* diseases, such as hypertension. Hypertension is sustained high blood pressure, which may eventually lead to heart attack, or the failure of such diverse organs as the kidneys, the eye, or the brain. Now there is a continuous spectrum of blood pressure variation in healthy people. In the disease case, the normal functioning of the sufferer's system (the circulation of the blood) is pushed by his lifestyle, his personality, or some other cause to the limits of tolerable variation, causing a gradual breakdown in his life functions. There is no natural discontinuity between the disease state and the healthy one; but we none the less use the language of disease, appropriately, because of the human effects of the condition.

In short then, the difference perspective is right and valuable in as much as it reminds us that there is a continuous spectrum between normal mental functioning and psychosis. Importantly for our purposes in this book, there are milder precursors of mania and depression, in elation and sadness, and milder precursors of schizophrenic thought, in unorthodox and divergent thinking. Indeed, the more extreme positions on the spectrum of mental life are, as we shall see, typical not just of malfunction—mental illness—but of the best in mental functioning—inspiration and creativity—too. Where the difference perspective, at least as articulated by Szasz, is wrong, is to thereby infer that mental disorder is not disease, and to draw huge conclusions about what should be done with schizophrenics from this. Psychosis is difference, but it is disease, too, because of its extremity and because of its human effects. In fact, the best definitions of disease actually stress that *all* disease is just difference; that is, departure along a spectrum a certain distance from a physiological norm. But not just any departure. More specifically, it is a departure that gives us therapeutic concern for the person. Thus the distinction between disease and health is partly a social construct, depending upon the amount of normal variation and suffering the society will tolerate. In previous epochs, large amounts of tooth decay and diarrhoea were considered normal. Now we intervene medically, because the threshold for therapeutic concern has been raised. The fact that where the line is drawn is somewhat arbitrary and culturally determined does not mean that the line should not be drawn at all.

In this chapter, then, I have introduced the major psychoses, and travelled from viewing them as diseases, to understanding them as expressions of human difference, and back to seeing them as diseases again. In fact they are both. Like photons in subatomic physics, which must be viewed now as particles, now as waves, we must see the psychoses now as discrete entities, now as spectra. This is because our concepts are provisional and porous, and we must not, at all costs, become slaves to them, or argue too long about semantics. This book will stress particularly the continuities between psychosis and normal mental life, and also between the different psychoses, but sometimes they are better seen as discrete entities. Both perspectives must be kept in mind. This will be especially apparent as I now consider where psychosis comes from: nature or nurture.

CHAPTER 2

From nature to nurture and back again

> Besides special tastes and habits, general intelligence, courage, good and bad tempers, etc. are certainly transmitted [in the blood].
>
> Charles Darwin, *The Descent of Man*
>
> They fuck you up, your mum and dad.
> They may not mean to, but they do.
> They fill you with the faults they had,
> And add some extra, just for you.
>
> Philip Larkin, *This Be the Verse*

Emil Kraepelin, as well as defining the major psychoses, began the study of heredity in mental disorder. That there might be an inherited factor had already been suspected for some time. Following the publication of Darwin's *Origin of Species* in 1859 and *Descent of Man* in 1871, there was widespread interest in heredity as the nineteenth century drew to a close. It was Darwin's cousin, Francis Galton, who had largely developed the idea that the human temperament was passed on through nature rather than nurture.

In his book *Hereditary Genius*, Galton attempted to assess the level of 'eminence' of people in the population. By eminence, he generally meant intelligence. There were, of course, no IQ tests at the time, and Galton had to rely on rating people's lifetime achievements. Galton produced a rudimentary bell curve for the distribution of human intelligence, which, although crude, was the precursor of modern psychometrics. He then observed from family pedigrees that eminent people tended to have eminent offspring, average people average offspring, and retarded people retarded offspring. There is clearly a problem here, in that eminent families provide a particular kind of nurture to their offspring, as well as genes, and in Victorian England that nurture was highly privileged. Galton never really dealt with this problem, although he did suspect that the study of twins would eventually provide a solution. He called his study of the inheritance of characteristics *eugenics*, a name which should make you shudder and realize immediately why there would be a reaction against it by the middle of the next century. However, this does not mean that Galton's observations were wrong.

The problem with applying Galton's methods to madness was that there existed no standard scheme for classifying mental disorders. As soon as Kraepelin had provided that, in his textbook of psychiatry which appeared between 1909 and 1915, the field was open. Kraepelin established an institute for the study of psychiatric genealogy, under Ernst Rüdin. In 1916, Rüdin published a study of the families of schizophrenics. He found that the risk of developing schizophrenia was several times higher than average in cases where a parent or sibling had the condition. The problem of shared environment was not really addressed, but Rüdin was convinced he had found evidence of heredity.

And here the story of eugenics begins to tarnish, before the really important studies had been done. Rüdin was involved in drafting the Nazi 'Law to Prevent Hereditarily Sick Offspring', enacted in 1933. Under this law, hundreds of thousands of psychotics were involuntarily sterilized and, later, almost 100 000 mentally ill people were exterminated. Involuntary sterilization was also widespread in less totalitarian countries, and the whole enterprise led to a widespread reaction against hereditary arguments and the science that underlay them.

◆

The human sciences, in the middle part of the twentieth century, were radically rewritten from a point of view that stressed nurture and learning at the expense of earlier ideas about nature and heredity. In general psychology, the movement known as behaviourism grew up. This had many strands, including a rejection of the subjective, introspective approach to mental life in favour of repeatable experiments, but the facet that concerns me here is that of its insistence on learning as the fundamental operation of the mind. The Russian scientist Pavlov had shown that by a simple, repeated learning operation, dogs could learn to respond to the sound of a bell as they normally do to food. There is no connection, in the natural environment of dogs, between bells and food. That they can learn to make one shows the malleability of their brains.

Behaviourist psychologists applied the same reasoning to all human behaviours, and showed that, indeed, people quickly learn to make many novel associations between things going on around them. The implications for the human temperament which the behaviourists drew from this were clear. People are products of learning, not nature, and by changing the stimuli you put into the system, you can change the person who comes out. The great behaviourists were not modest in their claims, which they extrapolated far beyond the rather limited experimental results they had actually achieved. J. B. Watson boasted that, given any child at an early age, he could turn him or her into any kind of adult society asked him to produce, purely by manipulating the environment. The difference from Galton's beliefs could not be more stark. Another behaviourist, B. F. Skinner, went even further. In his science-fiction novel *Walden Two*, the great psychologist fantasized a utopian world where conflict, disorder, and unhappiness had been abolished, by benign scientists optimizing the patterns of stimuli to which children (and to a lesser extent adults) were exposed. It is significant for what is to follow that, in this utopia, the family as the unit of nurture has been done away with, for this was the beginning of a period in which the family came to be seen as the root of most, if not all, mental evil.

Meanwhile, in anthropology, Bronislaw Malinowski and Margaret Mead were writing their famous descriptions of Pacific societies. Mead became particularly famous for her book *The Coming of Age in Samoa*. In this book, she purported to show that the emotional conflicts and social stresses associated with adolescence in the West are simply not found in Samoan society. Aggression and sexual jealousy are not part of their experience. The implication is clear; the patterns of violence, unhappiness, and evil we observe in our society (and this was a Europe that had recently seen the Great War) are not inevitable. They are a product of the particular patterns of learning that go on in our culture (this is the position in anthropology generally known as cultural relativism). If we can change those patterns of learning, the problem will go away. Unfortunately, it seems that Mead was factually wrong about Samoa. It is certainly the case that she knew what she wanted to show long before she got to the South Pacific, a fact that is perhaps not too surprising given the political circumstances in which she was working.

For its part, psychiatry came more and more under the influence of Freudian humanism in the middle part of the century. This makes sense at several levels. Kraepelinian biomedicine had lasted 30 years, and although it had classified the psychoses, it had not explained them, and it had failed to come up with any very effective treatments. (The drugs that are now the treatments of choice were all later discoveries.) Furthermore, it was implicitly tarnished by association with eugenics. The horror of the mid-century was seeing a world where people's fate depended on being put in some sweeping and hereditary category, such as Gypsy or Jew, psychotic or degenerate, and where people were processed by an awful, industrialized, efficient, dehumanizing machine. It was the image, in George Orwell's phrase, of 'a boot stamping on a human face forever'. Freudian humanism seemed a way of resisting this image. For Freud, the personality is uniquely shaped by early experience, particularly of sexual feelings, which are often repressed and percolate out in indirect and pathological ways. Freud himself was a refugee from Nazi Austria, and his work represented a defiant interest in the individual human, each of whom had a unique and important psychological journey. For Freudian analysis, no one is just another neurotic; everyone is a unique mass of conflicts, meanings, and experiences that make us who we are. Thus, whatever the huge scientific and political problems with Freud's specific theories, the overall enterprise is, in original essence, pro-person and pro-individual (and in this, as in other things, it contrasts sharply with behaviourism). This is the aspect that is brought out so starkly in D. M. Thomas' controversial and often brilliant novel about Freud and the holocaust, *The White Hotel*. The people who die in *The White Hotel*, whose psychodynamic fantasies we have explored in minute detail, are irreplaceable, and it appals us to see them dispatched like batches of objects.

This then, is the background to the mid-century triumph of nurture. Like many emancipatory movements, it started off as pro-people, but ultimately ossified into something as totalitarian as that which came before. This is because the influence of nurture can become the tyranny of nurture. It matters little to me if my mental problems come from my family, my culture, or my genes; what I want is the freedom to be able to change them. Behaviourism and cultural relativism and Freudian psychodynamics were all just as deterministic as Galtonian heredity. That is, they all eventually agreed that the human behaviour was strongly shaped by forces outside the subject's control, forces that might be too strong to overcome. Freudianism left just the escape rope of psychoanalysis, which was convenient for psychoanalysts, as this is extremely lucrative and gives them an extraordinary amount of power. Although the value of sympathetic psychological therapy for all kinds of problems is beyond question, the evidence that psychoanalysis in particular works is, alas, not very strong.

◆

Freud, unfortunately, had little to say about psychosis. His patients mainly had more minor conditions, and his method of psychoanalysis was developed to work on these. Indeed, he specifically claimed that he could do nothing for schizophrenics, since they were too far out of normal social interaction to develop the appropriate relationship with their analysts. However, there is no evidence that he thought psychotics were different in kind from other people with mental problems. Presumably, the early traumas were just greater, the repression of them more profound, the resulting distortions and sublimations more bizarre.

Later theorists in the broad Freudian tradition did extend his theories to psychosis. They followed his lead in implicating early experience, and in particular, traumatic interactions within the family, as the source of the problem. In 1948, the psychoanalyst Frieda Fromm-Reichmann coined the term *schizophrenogenic mother* to name and shame the suspected culprit in schizophrenia, and a few years later, the 'doublebind' theory advanced by Gregory Bateson and his colleagues came into prominence.

Bateson and his colleagues hypothesized that the symptoms of schizophrenia were generated in response to a problematic personal relationship, usually with the patient's mother. They suggested that the mothers of schizophrenics have mixed feelings towards their children. For example, a mother may, for some reason, feel some lack of affection or even hostility to the child, yet at the same time feel that a good mother *ought* to be affectionate, and wanting to be a good mother, she therefore wants to be affectionate. This can lead to the generation of mixed messages. They give the example of a young schizophrenic man who is pleased to see his mother on a hospital visit. He spontaneously puts his arm around her. She has some anxiety or reserve about physical intimacy, and she stiffens, so he withdraws his arm. But she also feels that a mother and son should be physically intimate, so when he withdraws his arm she says, 'Don't you love me anymore?' Now here the young man cannot win; he loves his mother, and wants her approval. At one level, she invites him to be physically intimate, while at another she rebukes him for doing so. Therefore if he advances on her, he incurs her displeasure; if he does not, he incurs her displeasure.

This young man is caught in what Bateson called the double-bind situation. Neither course of action will allow him to get what he wants. He is at a loss to know what to think. If he thinks he is wanted by his mother, his belief is contradicted by her deeds. If he thinks he is not wanted by his mother, his belief is contradicted by her words. The theory assumes that he is too young, dependent, and lacking in insight to see that the problem is not with him, but with his mother, and so the only way for him to make sense of the world is to make a mental split. He has to feel simultaneously close and distant to her, or he has to fail systematically to decode some of the signals being fed him by her. Only by misunderstanding reality in some way does his world make coherent sense.

The hypothesis was that double-bind patterns of communication, when repeated throughout childhood, lead individuals into thoughts or feelings that are ever more extremely split, and eventually into a schizophrenic breakdown. This is a strong nurture position, obviously; the causation of the psychosis lies entirely with childhood interactions, particularly with the mother. Of course, the symptoms of schizophrenia far exceed having contradictory beliefs about one's mother. We saw from the case of Mr Matthews in the Introduction that they can be vastly ramified and have nothing to do with one's parents at all. But here we see the Freudian lineage of Bateson's theory; it is assumed that a very wide range of human thoughts and feelings are really about one's mother and father, even if on the surface they appear to be about something quite different. There is another aspect to the double-bind theory which is important for what is to come. Under the hypothesis, the world that the schizophrenic is faced with is full of impossible contradictions; he is loved *and* he is not loved. The solution he comes up with has a certain logic to it—split your personality between two contradictory beliefs, or ignore one part of reality. Either way, it is not so much that he is crazy, but that the world is crazy and he is trying to find a way of making it make sense. This approach to psychosis—the psychotic is in some ways the sane one in an insane world—was a recurring theme of much anti-psychiatric writing, and was particularly developed in the work of the Scottish psychiatrist R. D. Laing.

Laing, who was very active in mental health care in Britain in the 1960s and 1970s, represented perhaps the furthest extreme of the swing of the pendulum away from nature and to nurture, so I will concentrate on him for the remainder of this section. A gifted writer who communicated through such forms as poetry, fiction, and dialogue, as well as the traditional case study, whose ideas ranged widely through his career, and whose non-medical, commune-style institution at Kingsley Hall was both famous and notorious, Laing became a major figure in 1960s counter-culture. Like Freud, he was able to capture the public imagination much more widely than his medical achievements would seem to justify. Also like Freud, he belonged to introspective humanism more than biomedical psychiatry; but he made far grander claims than Freud about his ability to cure psychosis.

Laing's early writing took up from the double-bind theory, and throughout his career he was concerned to show that schizophrenic beliefs make sense if one can only understand the situations that they are a response to. In his first book, *The Divided Self*, Laing explicitly opposed his view of psychosis to the Kraepelinian disease model. He quoted a description by Kraepelin of a schizophrenic patient. Laing argued that the utterances which Kraepelin had down as the random rambling of a diseased brain could equally be argued to be the legitimate non-cooperation of a young man who objected to being measured, tested, and stigmatized as a mental patient. Thus the problem is not so much inside the head of the patient; it is in the interaction between two people, one of whom society labels as sane, and one of whom it labels as mad.

Laing went on, in the book *Sanity, Madness and the Family*, which he co-authored with Aaron Esterson, to pursue the origin of psychotic 'symptoms' in patterns of family interaction. By now he was concerned to show that there is nothing intrinsically 'wrong' with the psychotic at all. Rather, there is an impossible or double-bind pattern of communication within the psychotic's family, and they deal with this by 'labelling' one member, perhaps the weakest, as insane. But the problem clearly lies in the patterns of communication between the individuals, not within any one of them. This work began Laing's propaganda to rehabilitate psychosis. The psychotic was seen as a lonely, almost heroic seer embattled by an impossible family determined to disempower him. Sometimes the suggestion seemed to be that capitalist society in general was crazy and full of contradictions, and that what we call psychosis was, at some level, the only sane response to it. In the preface to the 1965 edition of *The Divided Self* he wrote: 'The statesmen of the world who boast and threaten that they have Doomsday weapons are far more dangerous, and far more estranged from "reality" than many of the people on whom the label "psychotic" is affixed.'

From this point on, Laing's work celebrated the schizophrenic as a more whole and natural person than the sane one. Modern capitalist society demands obedient, one-dimensional minds, in which a fissure has been created between reason and experience, between the self and the other, and the psychotic is someone who has thrown off these shackles and gone on what Peter Sedgewick has called a 'radical trip' into personal growth. The consequences of this view for treatment (for the goal of psychiatry becomes supporting the tripper on his journey, not curing him as such), and the later changes in Laing's position, need not concern us here. What is of interest is the uncompromising attribution of causality for psychosis in Laing and his contemporaries. The fault, if fault there be, lies with nurture, not nature.

◆

By the 1970s, the arc towards nurture had reached its most visible extreme. Families, in particular, had been handed a lot of the blame. In addition to Bateson's double-bind theory for schizophrenia, it was widely believed that childhood autism was the result of 'cold' mothers, and explanations along similar lines were also in circulation for other disorders. Thus parents in the 1960s and 1970s often found themselves in the doubly awful situation of learning not just that they had a child with mental difficulties, but that it was also their fault. Philip Larkin captures the dismal end point of this period in characteristically gloomy fashion in the poem quoted in the epigram to this chapter. He finished that poem in 1971, 100 years after Darwin had published *The Descent of Man*. The arc had taken exactly a century.

However, even while the swing to nurture was most publicly visible, the evidence that would ultimately undermine it was quietly massing. A number of developments gradually pulled the consensus back towards the biomedical end of psychiatry. One of these was the discovery of drug treatments that were radically more effective than anything else the psychiatrist had ever had in his armoury.

In 1948, the Australian psychiatrist John Cade was performing experiments with guinea-pigs which involved injecting them with various compounds normally found in human urine, including lithium urate. Although he was researching manic-depression, he had no inkling that he was about to discover a drug, and in fact the lithium part of the compound was just there as a binding agent. However, Cade found that on application of the injection, the rodents became lethargic and relaxed. Further experimentation showed that the critical ingredient was not the urate, which Cade was interested in, but the lithium, since lithium carbonate and lithium citrate produced the same results. Cade had no idea what the mechanism of action of the lithium might be. This was medicine on the wing. None the less, he decided to make an imaginative leap and treat a manic patient, pausing first to take repeated doses of lithium salts himself as the most ethical way of establishing possible side-effects.

The results were dramatic. The first patient treated was a severely manic man, who had been hospitalized in a state of chronic excitement for 5 years. He was described as 'amiably restless, dirty, destructive, mischievous and interfering', and thought likely to remain on the back ward for the rest of his life. Within 3 weeks of the administration of lithium citrate, he was in a convalescent ward. In 2 months, he was at home, perfectly well, and he later returned to his old job. He was only kept in hospital as long as he was because of the scientific need to monitor the new treatment.

Cade's results were later confirmed by investigators all over the world, although they were not always so dramatic, and lithium was not effective in every case. None the less, they established lithium as the most effective treatment there had ever been for mania. What is more, lithium turned out to be effective both in pulling down the highs and pulling up the lows experienced by bipolar affective patients. How it does this is still not entirely clear, but it involves augmenting the effects of the brain chemical serotonin.

Lithium was not the most significant psychiatric drug to appear in the 1950s. The drugs reserpine and chlorpromazine were first used to treat schizophrenia in this decade. Once again, no one had much idea how or why they worked. The scientists involved were shooting in the dark. Reserpine was derived from a plant which traditional Hindu medicine ascribed with anti-insomniac and antipsychotic properties.

Chlorpromazine was synthesized initially as an antihistamine. Like may antihistamines, it turned out to have a sedative effect, and so it was used as a relaxant before surgery. Only by a little lateral thinking was the idea conceived of using it to treat psychotics, for whom, though no miracle cure, it has shown some benefit. Despite the serendipitous origins of these psychiatric drugs, the science behind them soon become a little clearer. The 1950s also saw the discovery of the first neurotransmitters. Neurotransmitters are the chemicals that the brain uses to pass messages between cells, a process that until this time had been obscure, but one that was clearly essential to brain functioning. The first neurotransmitters discovered (there are numerous different ones) were norepinephrine (also called noradrenaline) and serotonin. These belong to the chemical class known as monoamines. The identification of dopamine as a third member of this class was to follow quickly. The Swedish psychiatrist Arvid Carlsson was able to show that the drugs reserpine and chlorpromazine both inhibited the action of dopamine between brain cells, but in different ways. It seemed likely that this was related to their effects in damping down the symptoms of psychosis. (Reserpine and chlorpromazine also inhibit the action of other neurotransmitters, which we now know to be crucial for the maintenance of normal mood. This means that one of their side-effects can be depression.) Meanwhile, two other drugs, iproniazid and imipramine, had been discovered quite by chance to improve the mood, and were introduced as antidepressants. These drugs were subsequently shown to enhance the effects of all three monoamine neurotransmitters between brain cells. By the end of the 1950s, all five drugs—lithium, reserpine and chlorpromazine, iproniazid and imipramine—were in general psychiatric use, and, for four of them at least, it had been shown that they worked by controlling the actions of brain neurotransmitters. It is now known that the fifth, lithium, also works in this way. It was clear that psychosis has a biochemical basis, and that the neurotransmitters—the brain's messengers—are involved.

Now the finding that brain chemicals are implicated in mental illness does not destroy the nurture position. It could be the case, for example, that the imbalances of neurotransmitters that are found in psychosis are, in turn, caused by environmental stress or difficult relationships. However, the drug discoveries did take much of the sting out of the most radical nurture positions, such as that of Laing. It is difficult to argue that there is nothing at all wrong with the psychotic himself, but that he has just been 'labelled' as ill by a dysfunctional family situation, when drugs sort his difficulties out much better than removing him from his family. It is difficult to avoid the conclusion that there is something substantive, biological, internal to the psychotic, which is the basis of his condition. And this is the perspective of contemporary psychiatry.

As for the nature–nurture question, it is ultimately an empirical one, which can only be answered by careful study of the human population. Though the work of eugenicists such as Galton and Rüdin had fallen out of favour for a time, they had begun a systematic approach to human heredity, with proper statistical methods. This framework would eventually provide the answers to the question of the transmission of psychosis. I will turn to this work shortly; however, first it is worth running over very clearly what the opposing claims of the nature and nurture positions are, since nature and nurture, genes and environments, are not mutually exclusive things but, rather, interacting forces in the course of every life.

Nobody denies that psychosis can be brought about, in an immediate sense, by events in the social environment. Those who suffer a psychotic breakdown are more likely than average to have undergone major life stressors in the previous 6 months. Such stressors include bereavement, divorce, major family changes, or loss or change of job. As important as the stressors themselves are, the patient's attitudes and interpretations of those stressors also play a part, which is why 'talking therapies' can be so helpful in rehabilitation. However, although life events can be the immediate cause of the psychosis, they do not explain it, because they are not in themselves sufficient for psychosis to occur. Most, if not all, people will undergo major stressors several or many times in their lives. However, most people do not go mad. The difference between those who go mad and those who do not is not, by and large, the events they live through, but the way they are disposed to respond to those events. The events are the triggers, but all psychiatrists agree that there must something about the person that determines whether he goes mad in response to those triggers, copes, or reacts in some other way. Indeed, the triggers that push someone into a psychotic break can actually seem fairly minor from the perspective of someone who is more resistant. This is particularly true of relapses, since once there has been one episode of psychosis, repeats get easier and easier to evoke, and can be brought about by the slightest things. This is an effect known as kindling.

Where the nature and nurture positions part company is on the origin of the disposition to lapse into madness. For the nature position, it is a genetic liability. For the nurture position, the disposition is the result of patterns that have been set up in early experience, particularly of relationships within the family. Now there is some evidence that supports the influence of nurture. Parental death or divorce in childhood probably increases the risk of later affective disorder, and other difficult parenting relationships may also contribute to the probability of later breakdown. Complications during birth, and also birth during the winter, marginally increase the risk. These effects are relatively slight, however, compared to the genetic evidence I will review shortly. Most contemporary psychiatrists believe that early stresses of the nervous system can sensitize the individual to mental disorder. In other words, they can act as a kind of pre-trigger. However, they work on top of a basically inherited liability.

The strong claims of the nurture theorists are thus simply

not supported by evidence. It is quite easy to go looking at the parents of schizophrenics, as Laing and Esterson did, and find evidence of mixed feelings or mixed messages. However, it was never shown that you did *not* find such things if you look for them in the parents of non-schizophrenics. In short, the proper control was never done to show that something specific discriminated the parenting received by psychotic and that received by normal individuals. In the absence of such scientific evidence, laying the blame at the feet of the 'schizophrenogenic' mother was quite monstrous.

◆

It has long been known that psychosis runs in families. Indeed, this is one of the few points on which the most extreme nature positions and the most extreme nurture positions agree. The difference between them lies in how the familial pattern is to be accounted for. Is it a product of the genes, or a product of the patterns of social interaction which are perpetuated by families?

We can borrow a clear example of the family pattern from Kay Redfield Jamison's study, *Touched with Fire*. The example is the family of the English poet, Alfred Lord Tennyson. Because of the eminence of this family, we have particularly good biographical sketches over several generations, not least penned by the Tennysons themselves. These sketches allowed Professor Jamison to categorize the individuals as normal, bipolar manic-depressives, unipolar recurrent depressives, or as suffering from unspecified bouts of unstable mood and/or insanity.

As you can see from Fig. 2, the family pattern is strong indeed, with 8 of the 11 children of George Clayton Tennyson who survived to adulthood showing definitely identifiable symptoms of affective disorder. This observation is not an isolated one; dozens of pedigrees of schizophrenic and affectively ill families have been produced, and the familial pattern is unmistakable.

FROM NATURE TO NURTURE AND BACK AGAIN 51

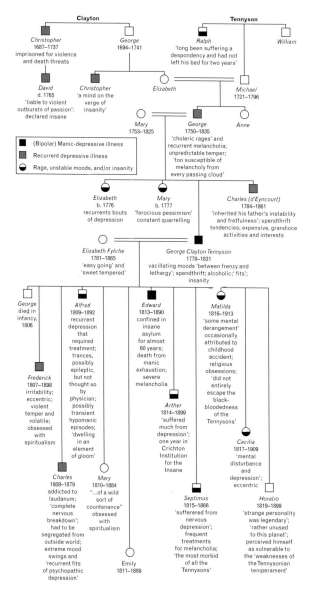

Figure 2 Partial family history of the Tennysons (reprinted with the permission of the Free Press, a division of Simon and Schuster, Inc., from Jamison, K. R. (1993). *Touched with Fire: Manic-depressive Illness and the Artistic Temperament.* Copyright © 1993 by Kay Redfield Jamison).

However, such observations are not unproblematic. For families such as the Tennysons, there are no properly objective psychiatric assessments or records. These things did not exist at the time. The investigator, who is not of course blind to the hypothesis under test, is therefore dependent upon his or her personal interpretation of whatever biographical material is available. We thus have to ask; do more rigorous contemporary studies bear the familial pattern out?

The answer is that they do. Modern studies of the distribution of mental illness, carried out by epidemiologists, proceed by choosing a given number of people, some with mental illness and some without. These individuals are called the probands. The investigators then measure, for each set of probands, the frequency of mental illness in their first-degree relatives. First-degree relatives are those one biological step away from the proband; mothers and fathers, brothers and sisters, sons and daughters.

For the affective psychoses, such studies show a moderate familial pattern. For the relatives of normal probands, the lifetime probability of affective illness is between 1 and 8 per cent, depending upon the exact criteria used in assessment. For the relatives of bipolar patients, this rate is between 9 and 41 per cent, 5–10 times higher than that for normal controls. For the relatives of major unipolar patients, the rate is between 6 and 20 per cent, 3–5 times higher than that for normal controls. Thus the disorder is familial, and the bipolar version appears to be more familial than the unipolar one. There is a further complication; the relatives of bipolar probands are more likely than average to be bipolar sufferers, but also more likely than average to be unipolar depressives. The converse does not hold, or not so strongly; the relatives of unipolar depressives are not especially likely to become bipolar. The best interpretation of this discrepancy is simply that the bipolar form is a more serious version of the disease, as we shall see in a later chapter.

Studies of schizophrenic probands suggest a similar conclusion, with the disorder just as familial as the affective psychoses. Irving Gottesman has usefully amalgamated the results of about 40 studies, carried out between 1920 and 1987, in his book *Schizophrenia Genesis*. The results are shown in Fig. 3. The figure shows clearly the increased risk as one's relationship to established schizophrenics increases. A person with no schizophrenic relatives has a lifetime risk of around 1 per cent. A person with one schizophrenic parent has a risk of around 6 per cent. A person with two schizophrenic parents, or a schizophrenic identical twin, has a risk of very nearly 50 per cent. Note that being the spouse of a schizophrenic slightly increases one's risk of schizophrenia, although not nearly as much as being a blood relative. This is obviously not to do with

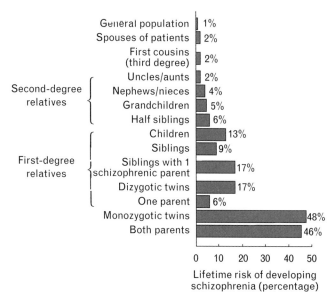

Figure 3 Risk of developing schizophrenia, by genetic relationship to another schizophrenic (from Gottesman, I. (1991). *Schizophrenia Genesis*. Copyright © 1991 by Irving I. Gottesman. Reprinted with the permission of W.H. Freeman and Company).

genes. The most likely interpretation is that this is an example of what biologists call *assortative mating*; that is to say, we seek out mates with similar characteristics to our own. Assortative mating is known to occur with manic-depressives too.

Now familial patterns, as I have already said, do not of themselves disentangle nature and nurture. They could be due to social interaction, or genes, or a combination of the two. Distinguishing between these possibilities is an important preoccupation of many recent psychiatric researchers, and they have used two main strategies to do it. The first is the study of twins, the second the study of adoptees.

Twins are interesting because they offer up a kind of natural experiment. There are two types of twins. *Monozygotic* twins are the product of a single fertilized egg, which subsequently splits. Thus they are genetically identical; the very same set of genes in two different individuals. Any differences between them cannot, in general, be due to nature. *Dizygotic* twins, by contrast, are the product of two separate fertilizations. Thus they share 50 per cent of their genes. They happen to cohabit *in utero*, but genetically, they are just normal brothers and sisters.

The logic of the twin studies is thus as follows. Most pairs of twins share the same nurture, whether they are monozygotic or dizygotic. Thus nurture is held constant. The difference is that monozygotic pairs also share all the same genes, whereas dizygotic pairs only share half of theirs. Thus any increase in similarity between monozygotic twins as compared to dizygotic ones is likely to be due to genetic effects. Roughly speaking, if the traits we are interested in are genetic, we might expect monozygotic twins to be twice as similar as dizygotic ones (although it is not actually quite as simple as this). The measure used by twin studies is the *concordance rate*. This is the probability that, if one twin has the disorder, the other also has it. If a disorder has a genetic basis, we should expect a higher concordance rate for monozygotic than for dizygotic pairs.

Studies of affective disorder reveal a concordance rate for monozygotic twins of between 33 and 93 per cent, average 65 per cent. For dizygotic twins, the rates are between 0 and 23 per cent, average 14 per cent. The picture is much the same for schizophrenia, as Fig. 3 shows. The average concordance rate for monozygotic twins is 48 per cent, whereas that for dizygotic twins is only 17 per cent. Having the same genes as someone with a psychosis seems to be the best predictor that you will develop it yourself.

The charge can always be brought against the twin studies that there are differences of nurture as well as nature at work. Parents may treat identical twins differently from non-identical twins. This could account for the differences in concordance rates. Such objections cannot be levelled at the other set of studies I will review; those of adoptees.

A large number of studies of mental illness in adopted children have been carried out over the past few decades. The logic of the adoption studies is simple. One merely has to discover whether the rates of mental illness in adoptees is more like the rates in their biological families, or more like the rates in their adoptive families, to effectively prise apart nature and nurture. The adoption situation is, if you like, another natural experiment.

One of the best known adoption studies looked at the prevalence of schizophrenia in the biological and adoptive families of schizophrenics identified from the Danish national registers. The probands were people who had been adopted soon after birth by families unrelated to their biological parents, and who had gone on to develop schizophrenia. The study found a prevalence rate of schizophrenia of around 6.4 per cent in the biological relatives of the probands. These relatives had played no part in the upbringing of the schizophrenic children, but their rate of schizophrenia was much higher than the population average. The adoptive families, who actually brought the children up, had a prevalence of 1.4 per cent, which is about the same as the 1 per cent found in the population at large. Thus, in this special case where one family provides the nature and another one provides the nurture, the tendency to schizophrenia is traceable to the family providing the nature, not the one providing the nuture.

There are more ingenious permutations of the twin and adoption-study paradigms, and they all seem to point in the same direction. Some studies have shown that being adopted *away* from a schizophrenic parent does reduce one's risk of developing the condition, but not by much. The risk is still much higher if one has a schizophrenic *biological* parent than if one does not. Furthermore, being adopted *into* a family that turns out to contain a schizophrenic, as happens from time to time, does not much increase one's risk. It seems that in the absence of the biological predisposition, exposure to a schizophrenic family is not sufficient to transmit the disorder. Finally, if one's parent is not schizophrenic, but is the identical twin of someone who is, one's risk of developing the disorder is just as high as if one's parent had it. That is to say, what matters is not whether one's parent shows the symptoms or not. If they have the genetic predisposition, as the healthy identical twin of a schizophrenic must do, the risk passes down.

Finally, what would be the clinching evidence on the nature/nurture question, if there were only more of it, combines both the twin and adoption strategies. This is the study of identical twins reared apart, one of whom is schizophrenic. Only 14 such cases have been described in the literature. Though this is a dangerously small sample for any generalization, the concordance rate amongst them is 64 per cent, which is not significantly different from the 48 per cent for identical twins reared together (although it is actually higher). The shared genotype seems to matter much more than the shared experience.

This, then, is the modern evidence for the position with which I began, with the likes of Darwin and Galton, Kraepelin and Rüdin. The findings are all strong evidence that hereditary factors are at work in psychosis. The nature position has much stronger scientific support than one based on nurture alone.

However, the picture is not simple. Many psychotics have no psychotic parent, whereas even having two psychotic parents does not guarantee that someone will suffer from psychosis. There must be a genetic basis, but it must be complex, and the genes must interact with other factors.

We can infer, then, that there are specific genes linked to psychosis. But all we know of the actual operation of genes is that they cause the production of particular proteins. How can the production of a protein lead to psychosis, in all its striking and variable forms? How many genes are involved, and what do they do in the working brain? And is the presence of these genes by itself a sufficient condition for madness to ensue? These questions are the topic of the next chapter.

CHAPTER 3

This taint of blood

> The root of the evil lies in the constitution itself, in the fatal weakening of families from generation to generation.
>
> Vincent Van Gogh

The evidence reviewed at the end of the previous chapter all pointed strongly to a role for hereditary factors in psychosis. This would seem to add madness to that long list of medical conditions for which particular genes are implicated. Our understanding of the genetics of diseases is currently advancing at a staggering pace. It is not just that we know the patterns of transmission that can occur; these have been understood at a theoretical level for over 100 years, ever since Gregor Mendel performed his famous experiments on the proportions of different flower types in hybrid peas. The current revolution is different. We now know the structure of DNA itself, and have cracked the code in which its instructions to body cells are written. We also have, admittedly indirect but none the less astounding, techniques for reading sequences of DNA from the chromosomes of living individuals. Thus, for hereditary diseases, we can now go hunting for the actual gene involved, and, having found it, establish how it does its work.

The classic example of this kind of research is the isolation of the cystic fibrosis gene. Cystic fibrosis is a moderately common disease (about 1 in 2000 births in Europe) which affects the developing body, and leaves its sufferers terribly vulnerable to infection and failure of the internal organs. Its seriousness means that, historically, few sufferers reached adulthood; even now, with better therapy, but no cure, the life expectancy of patients is just 28 years.

The hereditary nature of cystic fibrosis has long been obvious. It runs in families, but there are many individuals within affected families who are completely unaffected themselves. It is also possible for a sufferer to have both parents free of the disease (although there will often be a grandparent who had it). The best way of explaining this pattern, given that it is common, is to assume that parents can be 'silent' carriers. Let us assume that there are two alternative versions of the gene involved. A silent carrier is someone who has the disease version, and can therefore pass it on, but does not show the disease himself. As people have two copies of all their genes, the obvious inference is that carriers have one copy of the disease variant and one copy of the healthy one, and that two copies of the disease variant are required for the disease to manifest itself.

Such a variant is called *recessive*; to have its effect, two copies are required in the same person. Where a recessive variant sits alongside a different one, it keeps quiet. Recessive variants have a characteristic signature, as Mendel himself showed. When an affected individual mates with an unaffected one, the proportion of offspring affected is either none or exactly one-half, depending upon whether the *un*affected parent is a silent carrier. The remaining offspring are entirely healthy, but there may be a chance of the disease cropping up again somewhere down the line of their descendants.

The transmission of cystic fibrosis fits the recessive model very well. Researchers, armed with modern techniques for detecting particular sequences of DNA, could thus home in on the disease gene by looking for lengths of chromosome that were shared by just the affected individuals within a family. For cystic fibrosis, this search came to fruition in the late 1980s, when the gene was traced to a particular stretch of 4560 DNA bases on chromosome 7. This gene is now understood very well. It provides the blueprint for the manufacture of a protein that is involved in transporting chemicals across cell membranes. Any variation in the structure of this protein affects the biochemistry of the cell and has implications for the whole of the metabolism. It turns out that there are many different mutations to this gene in the human population—up to 300—but all of them cause cystic fibrosis, and all cystic fibrosis involves some mutation or other to this particular gene. A double dose of a mutant variant of this single gene is thus both the necessary and the sufficient condition for developing cystic fibrosis.

The cystic fibrosis gene is clearly very damaging when it leads to the disease. However, natural selection has not, as yet, removed it from the human gene pool. This is probably because having one copy of the disease variant does the bearer no harm, and might even do some good. This is the phenomenon known in genetics as *heterozygote advantage*; genetic variants stay around because having one copy, as around 1 in 23 of us does for cystic fibrosis, is adaptive. The unfortunate price of this is paid by the proportion of offspring of two carriers who get the double dose.

Cystic fibrosis is thus a paradigm case of a disease caused by a single gene, and a great success story for modern medical genetics. The isolation of the gene has led to greater understanding of the physiology of the disease, opened up the possibility of genetic testing, and pointed the way towards better therapies. With such advances going on elsewhere in medicine, the idea has naturally become widespread that a 'gene for schizophrenia' and a 'gene for manic-depression' will soon be able to be isolated in the same way.

By the early 1980s, several teams of geneticists in different countries were racing to isolate a gene for affective psychosis. The method of doing this is well described by Samuel Barondes in his book *Mood Genes*. Like the cystic fibrosis research, it involves studying large extended families that contain affected and unaffected individuals, and comparing genetic markers on their chromosomes. In 1987, the prestigious scientific journal *Nature* published a study by Janice

Egeland and her colleagues, using a large pedigree in which manic-depression was frequent from the Old Order Amish community of Pennsylvania. This study implicated a region of chromosome 11 in the transmission of the disorder. Frustratingly, though, the very same issue of the journal contained two reports from different populations which failed to find any linkage at all to chromosome 11, and further research on the Amish failed to confirm the earlier claim.

Meanwhile, a month after Egeland's paper, *Nature* published a study by Miron Baron and colleagues, implicating, for a population of Israelis, a locus on the X chromosome. The linkage to the X seemed plausible, as the trait in these families had the X chromosome's characteristic signature; it is never passed from father to son. (The X chromosome, one of the sex chromosomes, is never the same in father and son, as all men have only one copy, which comes from their mother.) However, here, too, subsequent research was ambiguous, and the same genes were not significantly implicated when the study population was expanded. Furthermore, in other families, transmission from father to son does occur.

Further studies are in progress in other populations, with an area on chromosome 18, for example, providing promising leads for a large pedigree of Costa Ricans. Weak associations have also been reported between schizophrenia and variants of several genes, although there is no single gene with a major effect. What is clear is that the neat cystic fibrosis story is unlikely to be repeated for psychosis. The rest of this chapter is devoted to explaining why this is the case.

♦

The idea that there could be a single gene involved in psychosis draws heavily on the view, discussed in Chapter 1, that psychosis is a discrete entity, an all or nothing, present or absent trait. In Chapter 1, I ended up endorsing the view that there was a continuum of mental functioning, with the psychoses occurring at the extremes of mood instability (for the affective psychoses) or divergent thinking (for schizophrenia). For both types of psychosis, we recognize a milder form (minor depression, or schizophrenia-like personality disorder) which goes some way down the road to the psychosis but stops short, as well as precursor traits which are found in basically healthy people.

Now if the presence or absence of psychosis depended upon the 'switch' of a single gene, the intermediate disorders would be difficult to account for. Do these people have the gene or not? This problem is particularly clear in studies of family pedigrees. Often, one finds descendants who do not have the full-blown psychosis but have some of the traits of their brothers and sisters. In Fig. 2 (p. 51), the pedigree of the Tennysons, we see that some of the individuals are manic-depressive, some have the depressive symptoms only, and some cannot definitely be diagnosed psychotic but have 'a wild sort of countenance' or some 'vulnerability to the Tennysonian spirit'. Now if there is a single gene, which of the Tennysons is carrying it? Just the manic-depressives? The manic-depressives and the depressives? The manic-depressives, the depressives, and those who stayed sane but were prone to moodiness? All of them? There is no parallel problem for cystic fibrosis, because one either has it, very obviously, or one is completely free from it.

If there were a single gene, its pattern of inheritance should, in principle, be rather clear. It should follow the simple Mendelian inheritance rules. If the psychosis variant is on the main body of the DNA and recessive, like the cystic fibrosis gene, then having one parent affected should lead to 50 per cent or 0 per cent or of the offspring being affected, depending on whether or not the unaffected parent has a copy. On the other hand, if the psychosis variant is dominant (which means that one copy is sufficient to cause the disease), then having one parent affected should produce the psychosis in 50 per cent or 100 per cent of the offspring, depending on how many copies the affected parent is carrying. Either way, the pattern should be fairly neat.

The real pattern for psychosis is not nearly so neat, and, statistically speaking, its transmission does not fit into a simple model of either one recessive or one dominant genetic variant. This is not a problem; geneticists can invoke the commonly observed phenomenon of *partial penetrance*. Partial penetrance is the name for the situation where a genetic variant does not have its usual effects on the organism in 100 per cent of cases. Thus, the transmission of the *gene* may conform to a simple Mendelian model, but the transmission of the *trait* is more complex, because the gene does not produce the trait in every case, and the trait is influenced in its development by environmental factors.

It is clear that psychosis genes are only partially penetrant. Remember that the concordance rates for identical twins are 48 per cent for schizophrenia and around 65 per cent for affective disorder. That is to say, when *exactly* the same genes go out into the same general environment and build a person, the result, as far as psychosis is concerned, is only the same 48–65 per cent of the time. However, the children of the unaffected twin are just as much at risk as the children of the affected one; the psychotic genotype is still there, albeit not expressed. The idea of partial penetrance can thus be appealed to save models based on a single gene. However, to make such an appeal is somewhat problematic. This is because the very methodology of 'single gene hunting' studies assumes a one-to-one relation between gene and trait. It is based on dividing the 'psychotic' family into affected and unaffected groups (already a binary category, note) and looking for the genetic variants that are in the affected members and not in the unaffected ones. The implication of this assumption is that having a particular gene is both the necessary and the sufficient condition for becoming psychotic (which is an all or nothing trait).

We know, however, that this is not so. As we have already seen from the study of identical twins, the same genotype only produces the same result 48–65 per cent of the time. Many psychologists believe that one's personality is the best predictor of psychosis risk, but even here the determinism is far from complete. Two people (such as two twins) can score identically on a range of psychometric tests, but have different outcomes; one can go into psychosis, while the other may have only mild personality difficulties, or he may be a successful and well-adjusted individual. Furthermore, psychotic breakdowns are very often triggered by environmental events. This is most obvious in the case of affective disorder, where bereavement and disappointment at work are often clearly implicated, but it may also apply to schizophrenia. Finally, anyone *can* have an episode of psychotic disorder, just as anyone *can* be sad or anxious. These states are potentially present in all of us. The difference lies in how easily they are evoked. A very un-moody individual might require really catastrophic life events to send him into depression; a very un-schizoid personality might require a massive stressor, like prolonged use of LSD, to send him into a schizophrenic break. But it could happen.

By contrast, some people's thresholds for these disorders are so low that those close to them always feel a breakdown may be coming, and when it does, the trigger events seem trivial. These are the people at high risk, and what they have inherited is the level of their threshold for evoking psychotic symptoms. As we shall see, this is systematically related to their personality. The genes, then, are not for psychosis itself, but for a predisposition or temperament, which, among other things, increases the chance of developing a psychosis. There are slips and gaps, as it were, between the predisposition and the disorder.

This view is known as the *temperamental threshold* position, and it has most recently been expounded by Gordon Claridge of the University of Oxford. It is a continuum position rather than a categorical one, but in rather a different sense from the argument I developed in Chapter 1. In Chapter 1, I argued that there was a continuum *of symptoms* that ran from the symptoms of psychosis, through the symptoms of more minor mental disorder, to the features of normal mental life. Within this continuum, there are no natural breaks, though psychiatrists may find it useful to impose diagnostic schemes. Claridge's position also acknowledges that symptoms come in varying degrees of severity. However, this is not the continuum he is interested in. His argument is that there is a continuum *of vulnerability* to psychotic breakdown. An individual's position on this continuum is inherited along with other aspects of personality. Being at the very vulnerable end of the continuum means two things. Firstly, the severity of triggering events that would be required to tip him over into psychotic breakdown is relatively small. Secondly, if he did break down, the severity of the breakdown would be very great. By contrast, someone lower on the continuum would take more to tip over into psychosis, and might have a milder episode if he were so tipped. Claridge distinguishes carefully between the temperamental threshold, which is part of the normal personality (everyone has a level of vulnerability, just as everyone has a blood pressure), and the psychotic disease itself, to which the vulnerable individual may or may not succumb.

The temperament threshold view allows for the great heterogeneity of psychotic episodes. In some cases, an individual at intrinsically low risk may be pushed over the brink by traumatic life events, while, in others, the individual is so prone to negative or strange thinking that very little is required to set it on. This spectrum accords with common experience, and is well supported by psychiatric and psychological observations. It also suggests strongly that we should think of the transmission of psychosis as more akin to a continuous variable such as height or weight, than to a single-gene, present or absent, disorder such as cystic fibrosis.

◆

Human height is under partial genetic control. There is, of course, an environmental aspect as well, as average height varies strongly with nutritional conditions, and has been increasing for the past couple of centuries. However, one's place on the height scale relative to the average of one's generation is strongly related to the place of one's parents relative to the average of their generation.

The heights of any population are distributed in a classic 'bell curve' with no natural breaks, most people falling in the middle, and fewer and fewer people as one moves out towards the extremes (Fig. 4). If a very tall person and a very short one produce children, we do not find 50 per cent of these children very tall and 50 per cent very short, as we might expect if height was controlled by a single gene. Rather, we find a range of heights in the offspring, which are generally close to the *average* of the heights of the two parents. Two very tall people will on average produce tall children, whereas two very short people will, on average, produce short children, and one short and one tall will, on average, produce heights in between.

This pattern is explained by the fact that height is under the partial control of many independent genes, each of which adds a little to the overall position of the person on the human height scale. Such a trait is called *polygenic*. Given that genes are, by and large, inherited 50/50 from each parent, the child will tend to share 50 per cent of these height loci with father and 50 per cent with mother, and so the genetic forces acting on his development will be a kind of average of the genetic

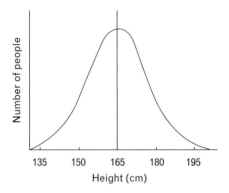

Figure 4 The distribution of height in a human population.

forces that acted on his parents. Note that because of the lottery of inheritance, there is a chance of two non-tall parents producing a tall child, or vice versa. This is because, in two average parents, there are both genes that move them up the height spectrum, and genes that move them down, and their height is the result of the two. The child, metaphorically, puts both his parents' genes into a bag and draws half, and he might happen to draw the half that push towards tallness, not the other half. By the law of averages, this will happen less often than the case in which he turns out to be the same height as them, but it can happen.

Human personality, as we saw in Chapter 1, is measured by the subdiscipline of psychometrics. According to psychometric theory, personality varies continuously along several dimensions in a bell-shaped distribution. Thus, just as one could make a plot of the height and weight of a human population, which would show most people in the middle and fewer and fewer out to the edges, one could plot all the personalities in a population on dimensions such as extraversion or neuroticism. The exact number of dimensions one would need for such a plot is unimportant to our present purposes; several current theories use five, but we can stick to two by way of example.

It turns out that one's position in the personality distribution of one's generation is strongly related to the position of one's parents in the personality distribution of their generation, just as was the case for height. The strongest evidence that this is a genetic effect comes from the same sources as the evidence that there are genes for vulnerability to psychosis: twin and adoption studies. Such studies show a high concordance between biological relatives, and particularly between monozygotic twins, *whether or not* they are brought up together, both on 'soft' questionnaire tests and on reactivity experiments. As for height, a polygenic model turns out to be appropriate for the inheritance of personality.

So what exactly is the personality trait that predicts one's vulnerability to psychosis? This is an area of ongoing research and discussion. The most influential early attempt to formulate it can be found in the work of Hans Eysenck. Eysenck developed personality questionnaires which could be used to place everybody in the population at some point on two now-familiar axes: introversion–extraversion (the E axis) and stability–neuroticism (the N axis). Eysenck later added a third axis, which he called psychoticism (P). The individual with a high psychoticism score was typically 'solitary, troublesome, cruel, lacking in empathy, hostile, sensation seeking, and liking odd or unusual things', and was hypothesized to have a high probability of psychotic breakdown. Indeed, Eysenck claimed that groups of schizophrenics and of bipolar psychotics had much higher P scores than normal subjects. This, then, appeared to be the dimension that determined the psychosis-proneness of the personality. Unfortunately, there are several serious problems with the psychoticism axis.

For one thing, the items on the questionnaire did not prove very reliable or internally consistent, although this was remedied partly by developing improved scales. Secondly, P seems neither to be specific to psychosis nor strongly predictive of it. The groups who are most obviously high scorers in terms of P are psychopaths, criminals, the violent, sadomasochists, and delinquents. Professors Jean and Loren Chapman of the University of Wisconsin administered the P scale to a large cohort of college students, whom they re-interviewed 10 years later to see what their P scores would predict. It turned out that high P was strongly predictive of later criminality and substance abuse, but not predictive of later psychotic breakdown. For this reason, critics have argued that Eysenck's psychoticism scale is misnamed; it is a measure of antisocial tendencies, which is interesting in its own right, but not strongly related to psychosis.

A third and related problem with the P scale is that it is not entirely independent of neuroticism, particularly in psychiatric populations. High scores on the two scales tend to go together. In fact, high N, rather than anything else, is the best predictor of mental disorder in general. Eysenck's view was that the neuroticism score would predict the occurrence of neuroses, and the psychoticism score, the occurrence of psychoses. This view now seems quaint. We now think of neurotic and psychotic disorders as related on a continuum, not as different in kind. Perhaps, then, high N should predict psychosis. After all, neuroticism was developed as a 'general measure of emotionality', 'mood swings', and 'the lability of the nervous system'. This seems to describe the psychotic, particularly the affective psychotic, pretty well.

Unfortunately, this does not quite work either. N can be quite labile: in affective psychotics, N scores are higher when the patient is manic or depressed than when they are in remission. Overall, high N is a good predictor of mental problems in general, but it does not specifically discriminate the liability to psychosis. This is no great surprise if the liability of psychosis is seen as continuous with the liability to more minor psychological problems, but it would be useful to have a sharper instrument of investigation.

Among the most promising current candidates for such an instrument are the various scales developed by the Chapmans in Wisconsin. These questionnaires investigate the occurrence of experiences in normal life that bear some resemblance to psychotic symptoms. Thus, the respondent is asked to respond to a large number of items that tap into whether he has had anything like delusions, hallucinations, strange emotions, perceptual distortions, and so on. The idea is that those with a predisposition to psychosis may have some of the psychological characteristics of the disease in their normal, pre-breakdown patterns of thought.

The Chapmans' scales were administered to a large number of healthy individuals who, once again, were re-interviewed 10 years later. High scores on two of the scales in particular (the so-called 'perceptual aberration' and 'magical ideation' scales) were strongly predictive of later psychosis. They were also predictive of the occurrence of psychosis in other family members. These scales clearly tap into the psychological factors that make up the inherited predisposition to psychosis. What is more, scores on the scales have been shown to correlate with other, 'harder' measures of nervous system function, such as brainwave patterns, the ability to discriminate simultaneous stimuli, eye tracking movements, and the functional differences between the two brain hemispheres. All these measures are known to show differences between psychotics and normal subjects.

The Chapmans' scales are not yet perfect as measures of the personality dimension underlying psychosis risk. The main problem is that scores on them do not distribute in a bell curve when the whole population is sampled. Instead, there are large numbers of people who score zero or nearly zero, and a few deviants. This is in contrast to the Eysenck scales, which provide a continuous bell-shaped curve, which is what we should expect when a real polygenic continuum is measured. This has come to pass because the Eysenck scales came out of the psychology of normal personality, with its interest in capturing the variation in the whole population. The Chapmans' scales, by contrast, were born in the clinic, with a concern to detect just those individuals with a high risk of psychiatric illness. The solution may be new scales that combine the Eysenckian and clinical approaches, and such scales are being developed, most notably under the general title 'schizotypy' by Gordon Claridge and his colleagues in Oxford. In the meantime, the general lesson of psychometrics is fairly clear: there are temperamental traits that predict one's liability to psychosis, and it is these traits, inherited genetically, that run down the generations of psychotic families.

The Chapmans' scales were constructed with schizophrenia in mind, but it turns out that scores on them also predict the occurrence of affective psychosis. This raises important questions: are the traits that make one liable to schizophrenia the same ones that make one liable to mania and depression? Are the same genes involved? Or, alternatively, are they two different dimensions, with different sets of genes, but our techniques are not yet sophisticated enough to discriminate between them? I will return to this question at the end of the chapter.

For now, let us assume that there is a single dimension of psychoticism, which measures the liability of an individual's nervous system to psychotic breakdown of whatever type. Although I use Eysenck's term, psychoticism, for convenience, the trait I am referring to is not exactly that measured by Eysenck's P dimension, for the reasons I have discussed. It is also important to keep in mind the fundamental difference between *psychosis*, the actual condition of madness, and *psychoticism*, the inherited personality trait that makes one vulnerable to madness. The latter may or may not get converted into the former, and indeed the latter is present in many perfectly sane people.

My model of the human population, then, is as set out in Fig. 5. There is a continuum of personality variation in psychoticism. The further you go up the psychoticism scale, the greater the risk of psychosis. Note, though, that appropriate triggers are required, and even the individuals highest on the psychoticism scale are not guaranteed to become ill. Indeed, the identical twin concordance rates suggest that only around half will do so. None the less, position on the scale is, as for other personality traits, fairly heritable. This inheritance is polygenic. Many different genes affect the trait, and so the position of offspring will, on average, reflect the average of the positions of their two parents, with some variation due to the genetic lottery.

A model such as this accounts for a large number of common findings. Firstly, it predicts that no single gene will exist that is present in all psychotics and absent in anyone else. This is indeed the conclusion that the authors of the largely inconclusive 'gene hunting' studies have reached.

Secondly, the model sits comfortably with the finding that psychotics are not categorically different from non-psychotics.

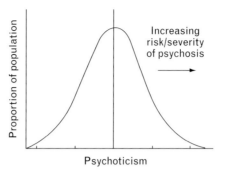

Figure 5 The distribution of psychoticism, and its relationship to psychosis risk.

There are intermediate forms of symptoms, conditions, and tendencies within the non-psychotic population. This is particularly obvious in the study of family pedigrees. If there were a single gene, we would expect to find, as in families with cystic fibrosis, a binary split between those members who have the problem (and the gene) and those who do not. Instead, in psychotic families, we commonly see the pattern we saw with the Tennysons. They had psychiatric problems of varying severity; some full-blown psychoses; some milder personality disorders, and some people who were well but rather extreme in personality, in the direction of the family taint. Pedigree studies for both schizophrenia and affective disorder have repeatedly shown that if you have a relative with a psychosis, your chance of developing that psychosis is still quite low. Your chance of developing a milder personality disorder, or neurosis, of related type, is much higher. This would not be the prediction if there were a 'have it or not' gene. It *is* the prediction of the idea that you cluster near your relatives on a continuous scale of temperament. Further support for this point comes from the fact that many of the psychometric peculiarities of psychotics, such as eye-movement irregularities, are found in their *unaffected* relatives as well. The difference between the well and the ill siblings, then, is probably not presence or absence of a biological vulnerability, but either the strength of that vulnerability or the presence of an environmental trigger to turn it into psychosis.

Thirdly, the continuous polygenic model accounts quite comfortably for the fact that 90 per cent of schizophrenics, and a similar proportion of affective psychotics, have neither parent with the disorder. This could, admittedly, also be explained by a single recessive gene, of which the parents were silent carriers. However, it follows quite naturally from a polygenic situation in which a child can be, due to the shuffling of the genetic pack, some way above or below his parents on the scales of temperament.

Finally, the polygenic model accounts for what are known as severity effects. The siblings and children of schizophrenics whose psychosis is mild, remitting, or clearly triggered by an external event have been shown to be at lower risk than the relatives of those whose psychosis is more severe. This is also true of identical twins. Where one twin's schizophrenic episode is mild (less than 2 years in hospital) the concordance rate drops to 27 per cent, around half its usual level. We can assume that the more severe cases are simply further out on the continuum than the milder ones, and so their relatives will also tend to be further out. A similar argument can be made for affective disorder. The relatives of full-blown manic-depressives are at greater risk of suffering any affective disorder than the relatives of those with unipolar depression only. If we assume that manic-depression is a more severe form of mood disorder, this could be counted as a severity effect. There is no way of explaining severity effects within a single gene model; either my parents have the gene or not, and they either pass it to me or not. In a polygenic model, severity effects are to be expected, for the further my parents are along the scale, the further along I will tend, on average, to find myself, and that means a lower threshold and greater vulnerability. Inheritance in psychosis is certainly not simple, with details of the pattern still to be worked out, but it is quite clear that multiple genes are involved.

♦

Let us grant, then, that there are many different genes, the variants of which contribute to variation along the dimensions of human personality. What do these genes actually do, at the biochemical level? The detailed answer to this question is not yet entirely clear, but the general picture is that they code for the development of brain structures and the production of certain brain chemicals, such as neurotransmitters. It is time to understand a little more about what these structures and chemicals are, and how they could affect a person's disposition. To do this we have to make a brief foray into the world of the working brain.

The brain consists of a vast web of billions of specialized cells. Crucial amongst these are the neurons that, crudely speaking, send and receive messages. The incoming messages come from the body—the sense organs and nerve endings—and from other brain cells; and the outgoing ones go to other brain cells, to muscles, and, via the pituitary gland, to the hormonal system that controls many aspects of the body's state. The brain is very different from a computer, but it works by a similar principle; each cell is relatively simple, in terms of the computation it performs (although a neuron is a lot more sophisticated than the logic gates in your computer). The complexity of the processing that can be done comes from the fact that a huge number of cells are wired up in a very specific way.

The brain consists of a large number of partly autonomous neural circuits, dedicated to particular functions. Their partial autonomy can be deduced from the fact that when people suffer very localized brain injuries, rather than a general intellectual impairment, they usually lose ability in a highly specific area, such as the perception of faces, the perception of movement, grammatical language, memory, or emotional tone. The implied functional specialization is confirmed by studies using the PET scanner, which show that specific areas of the brain become more active when particular types of task—emotional, memorial, executive, perceptual—are undertaken.

The neurons that make up these circuits consist of a long wire (the axon), and, at the other end, thousands of small twiggy endings (the dendrites) that attach to adjacent cells. Thus each cell is multiply connected; hundreds of dendrites from other cells terminate all over its surface, and its dendrites, in turn, terminate on hundreds of other cells. The cell fires messages, in the form of small electrical charges, which propagate down to the end of its dendrites. Its firing of such a message (known as an action potential) is triggered in turn by the sum of the inputs, like, simplistically, a semaphore team in which a man waves his flag when enough of the people immediately behind him wave theirs. Action potentials can fire off at a rate of dozens per second.

From all this, it might appear that the neural circuits are like a telephone network, with each point being joined up by the dendritic cables, and electrical impulses passing between them. This view is wrong in one important respect, which was demonstrated by the great Spanish neurologist Santiago Ramón y Cajal, who received recognition for his work from the Nobel committee in 1906. In fact, there is a gap between the ending of one neuron and the beginning of the next, over which electrical impulses cannot pass directly. This gap is called the synapse, and for the message to cross the synapse, it must be put into a different medium, this time of wet chemistry rather than electricity. It is rather like, to stretch the metaphor, a telegraph network where the telegraph cables cannot cross numerous small rivers. Every time a river is reached, a man must take the message, leap into a boat, row across, and re-enter the message at the telegraph station on the other side. This is where neurotransmitters come in. They are the boats.

An action potential, arriving at the cell membrane (river bank), triggers the release of many molecules of neurotransmitter from small vesicles in which they are stored. The neurotransmitter molecules flood out into the cleft of the synapse, many of them landing on the surface of the next cell (the other bank of the river). On that surface are specialized molecules, called receptors, whose shape is such that the neurotransmitters, but no other molecules, bind to them as a key to a lock, causing them to change their state in the process. The receptors are of different types. Some, when bound by a transmitter, cause a channel to open through the cell membrane, which admits electrically charged ions. This changes the electrical potential of that part of the cell, and thus contributes towards the production of an action potential. Other receptors do not open a channel, and therefore do not directly contribute to the firing of a new action potential, but instead cause longer-term chemical changes on the inside of the neuron, which can include making more neurotransmitter or neuroreceptor chemicals. The neurotransmitter, its job done, is either broken down in the synapse by enzymes, or transported away by special chemicals, to be reabsorbed.

There are dozens of different neurotransmitters in the human brain, although each synapse is specialized for a particular one, since the receptors on the post-synaptic neuron have to match the structure of the particular transmitter coming from the pre-synaptic neuron. There are also many different types of receptor for each transmitter, with different reactive properties. The populations of cells using different transmitters and receptor types have identifiable geographical locations in the brain, which is presumably related to the functional specialization of the circuitry. It is easy to see that either varying the structure of a specific circuit, or varying the level of production of the transmitters relevant to that circuit, would vary the behavioural disposition of the person. For example, Parkinson's disease can be shown to involve a deficit of a neurotransmitter, dopamine, in certain mid-brain systems that are involved in bodily movement. A deficiency in the transmitter system means that control is lost, and the patient suffers small involuntary movements, or tremors. Treatment with L-dopa, which is a chemical precursor of dopamine, helps to suppress this.

The success of treatments like L-dopa might lead us to adopt a simple notion of 'one circuit–one transmitter–one function'. That is to say, dopamine is the transmitter controlling the movement system, therefore lack of movement control means an impaired circuit or too little dopamine, and people with fine movement control are just people whose circuitry or level of dopamine is better developed. Such reasoning has been applied to many different functions over the past couple of decades.

One example is the relationship between mood and the neurotransmitters norepinephrine and serotonin. The link between these transmitters (which, together with dopamine and acetylcholine, are called the monoamines) and disturbances of mood became apparent in the early 1950s, when it was discovered that the drug reserpine, which can cause depression, depletes their level in the brain. Furthermore, iproniazid and imipramine, which were discovered by chance to work as antidepressants, turned out to do their work by inhibiting the chemical breakdown of monoamines in the synapse. This led to a general supposition that it was the level of monoamine function that determined the person's mood, and subsequent generations of antidepressant medicines have worked by enhancing this level in ever more targeted ways.

But which of the monoamines is the critical one in human mood? Norepinephrine seemed the early favourite. Levels of norepinephrine by-products in the urine and cerebrospinal fluid of many depressed patients are abnormally low, and postmortem studies of the brains of suicide victims show an increase in the density of norepinephrine receptor molecules. Now one might assume this meant that norepinephrine had been particularly abundant just prior to death, but in fact this is not so. Receptor levels vary in see-saw fashion, rather than in lock–step, with neurotransmitter levels in the brain. That is to say, when transmitter levels are low, the brain manufactures more of the receptor to compensate, and the other way around. The high levels of receptor therefore indicate a low level of transmitter.

More recently, much of the smart betting in the depression stakes has moved from norepinephrine to serotonin. This is mainly due to the phenomenal success of the drug Prozac (fluoxetine), and its cousins. These drugs enhance the level of serotonin activity in the brain, and they greatly improve positive mood in most people. Furthermore, some, but not all, depressed patients show reduced levels of chemicals related to serotonin in the blood and cerebrospinal fluid, and, just as for norepinephrine, post-mortem evidence suggests a high level of serotonin receptors in certain parts of the brains of suicide victims.

Which of these chemicals, serotonin or norepinephrine, is the important one? The answer is that we should change the question a little. Rather than thinking of 'one chemical–one function', we should remember that the brain is, from a chemical point of view, a complex, self-regulated web of interconnected chain reactions. It goes wrong when the balance of these is out of kilter, and the imbalance can probably start at many different points on the web, and have knock-on effects at many others. For one thing, the monoamines are highly interdependent. Dopamine and norepinephrine activity is quickly stimulated by stressful or arousing situations; serotonin activity follows more slowly with repeated exposure to such situations, and seems to regulate the functioning of the other two (although there is also a regulatory link from norepinephrine back to serotonin). The monoamines, and the neural circuits they serve, are thus highly interlinked. Depression can be caused by any drug that inhibits the actions of any one of them (excluding acetylcholine, which functionally seems quite distinct), and any drug that enhances monoamine function can have an effect on mood. Indeed, the very latest antidepressants, such as venlafaxine, target both norepinephrine and serotonin levels.

Similarly, the activity of dopamine is usually argued to be abnormal in schizophrenia, but it does not operate independently of serotonin in this role. Antipsychotic drugs affect the serotonergic as well as the dopaminergic system, both because of the interlinkage of the two systems and because the monoamines are chemically so similar. Some antipsychotics are thought to have their main effects through operation on serotonin rather than dopamine, just as some antidepressants have theirs through norepinephrine rather than serotonin. Furthermore, mania also involves excessive dopaminergic activity, though affective disorder in general is thought to start from a serotonin or norepinephrine imbalance. This means that some of the antipsychotic drugs suitable for schizophrenia can also help with mania.

As another example of interdependence amongst the monoamines, one of the most common side effects of serotonin-based antidepressants is a small tremor, a bit like that seen in Parkinson's disease. These drugs flood the cleft of serotonergic synapses with extra serotonin. Serotonin is not implicated in motor control, but dopamine is. Presumably the tremor occurs because the serotonin surge has a knock-on, suppressive effect on the dopamine systems somewhere deep in the mid-brain. The transmitter circuits are thus not functionally independent but all connected up.

We should also remember that the neurotransmitters themselves depend for their functioning on a whole web of other chemicals. There are many different types of receptor for each monoamine, and these different receptors are concentrated in different brain regions and modulate different functions. The receptors include so-called autoreceptors, which detect how much of a particular transmitter a cell is producing and cause new supplies to be manufactured accordingly. There are also the enzymes that break the transmitters up, other chemicals that transport used neurotransmitter molecules away from the cell membrane to be reabsorbed, and, finally, various proteins involved in the transport and manufacture of all these substances. For a person to be responsive to the world in the normal way, there must be a kind of equilibrium between all these chemicals. The transmitters must be produced at an appropriate level in response to brain activity, the receptor chemicals must be produced at a rate appropriate to detecting the level of neurotransmitters present, and the chemicals for removal and deactivation must be sufficiently plentiful to maintain the basic level, not too plentiful (which would deplete transmitter levels) nor too scarce (which would lead to too much transmitter building up). Thus, in psychosis, what we are seeing is probably not as simple as an excess of some transmitter or other, but rather a kind of disequilibrium in the relative levels of transmitters, receptors, and, perhaps, transporters. The best evidence for this view comes from the way modern antidepressant medications work.

Both the latest antidepressants of the Prozac family, and the earlier monoamine oxidase inhibitors, work by blocking the destruction of used neurotransmitters at the synapse. The Prozac family, for example, deactivates the biochemical pump that transports serotonin away from the synapse once it has done its job. Thus the drugs greatly increase the levels of serotonin in the synapse. Now this increase occurs within hours of the patient beginning medication. However, the therapeutic effects on the patient's mood do not generally occur for a few weeks, so it is not the raised level of serotonin *per se* that accounts for the improvement. It is more likely that in depression there has been some kind of skew between the levels of transmitter and receptor. This may account for the depressive feelings of lowness and anxiety in the absence of any external cause, and of lack of pleasure in the things that once brought pleasure rushing in. The transmitters may still surge in response to these things; but the levels of receptors seem for some reason to be inappropriate to register that surge as pleasure. Flooding the synapses with extra transmitter, as the drugs do, eventually causes readjustment in the level of receptor functioning, and this allows normal reactivity in the relevant brain centres to be resumed.

The point of this excursus has been to establish two things. The first is that a person's disposition, his way of reacting to things in the world, depends on the neural circuits he has and the levels of the chemicals within them: neurotransmitters, receptors, transporters, and so on. The second is that these circuits and chemicals function in a massively interdependent way. These insights apply not just to psychiatric disorder, but to the normal personality as well. Personality variation in the human population is related to variation in these systems between people.

Evidence that this is the case is beginning to emerge. For example, it has been shown through drug interventions that the reactivity of a person's dopamine system is related to the extent to which he will seek out new and stimulating experiences. This trait, known as 'sensation seeking' or 'novelty seeking', is a stable personality axis and has a strong hereditary component. How the heredity works was a matter of speculation until only 5 years ago, when reports of a series of studies combining psychometrics and molecular genetics began to appear.

In January 1996, an Israeli research team led by Richard Ebstein, published results of their ongoing investigation into the human genes that code for the production of dopamine-receptor chemicals. One of these genes makes the fourth-discovered type of dopamine receptor, D4, and is hence known as *D4DR*. This gene is interesting because it contains what is known as a hypervariable sequence. That is to say, there is a stretch of DNA within it, 48 base pairs long, which repeats itself, and the number of times it does so varies greatly from person to person. Most people have 4 or 7 repeats, but there are variants in circulation with as few as 2 and as many as 11 copies of the sequence. This variation in the gene is enough, biochemically, to change the effectiveness of the receptor, and so we would suspect a possible relationship between the length of the gene and the person's score on personality traits related to dopaminergic activity.

Ebstein and his team found that in their sample there was indeed a correlation between the number of repeats in the *D4DR* gene and the trait of sensation seeking, as measured by a psychometric questionnaire. In general, the more repeats a subject had in *D4DR*, the higher they scored on measures of openness to novelty, sensation-seeking, or positive emotionality. These traits strongly affect the life choices people make, in terms of career, relationship patterns, sexual habits, and leisure activities.

Ebstein's results were simultaneously replicated by a group led by Jonathan Benjamin on a completely different population in the USA. Other teams raced to replicate the correlation in other locations, but several of them have notably failed to do so. The state of play at the moment is unclear, with the negative reports rather outweighing the positive ones. The correlation may yet prove valid, but there are various methodological difficulties to be overcome, and the effect of *D4DR* on personality is at best a weak one.

A parallel and equally suggestive result has been obtained for genes related to the serotonin system. A gene has been isolated, the main function of which is to regulate another gene, which in turn produces the protein that transports serotonin away from the synapse (this is the protein disabled by Prozac). The first gene exists in two forms, a long and a short one. About one-third of us have two copies of the long variant, with everyone else having at least one copy of the shorter one. Given the implication of serotonin signalling in mood, we might expect some relationship between this gene and the emotional aspects of personality.

The first study to provide a correlation of genetic make-up with psychometric scores once again proved encouraging. People with the short form of the gene scored significantly higher on the neuroticism axis than those with two copies of the long form. High neuroticism scores predict, in life, anxiety, mood fluctuations, and the risk of psychiatric disorder. This is exactly what we might expect a disruption of serotonergic systems to do, from what we know about manipulating serotonin through drugs.

As with *D4DR*, results concerning the serotonin transporter gene have not been consistent. There have been several negative replications of the correlation with neuroticism. However, there is some evidence linking possession of the short form of the gene with susceptibility to depression. The effect is once again a small one at best, which may explain the failure of several studies to detect it, and methodological refinements are called for. None the less, this may well be one of the genes that moves the human temperament towards the liability to mental illness.

Research in this area is moving forward at great pace, and the precise importance of *D4DR* and the serotonin transporter is yet to be established. However, the positive results at least suggest a plausible picture of how personality is inherited—gene variants lead to individual differences in the reactivity of different neurochemical circuits. The two genes described must have, at best, a very small effect. *D4DR*, even in the original positive studies, accounted for only about one-tenth of the variation in sensation seeking that could be attributed to heredity. However, this is no surprise. A polygenic model has long been believed to be right for the major personality axes, and indeed we can see why this must be the case. There are genes that code for each and every transmitter, each and every receptor, each and every transporter, and other genes that turn those productive genes on and off. Any genetic variant that affects the rate or efficiency of any one of these biological steps will affect the relevant personality factor. Furthermore, in view of the great interdependence of brain chemicals, it should not surprise us if there are dozens, or even hundreds, of genes with a potential effect on a given trait.

However, it might be useful to recognize two different classes of genetic variants that will affect the personality. Some mutations, by coding the instructions for developing cells to divide and grow, affect the wiring of the brain permanently from infancy. They will affect neurotransmitter function, in that the circuits used by a particular transmitter may not be there at all, may be atrophied, or may be abnormally connected up. These kinds of mutations contrast with those of genes such as *D4DR*, which operate in a more exclusively chemical way. They affect the regulation of the reactions in a particular neurotransmitter circuit, but not, it seems, the development of the circuit in the first place. The problems associated with these latter mutations will be relatively reversible by means of drugs, whereas those of the former type will be rather more permanent. The present evidence points to more of the former, structural type of changes in schizophrenia, and more of the latter, chemical type of changes in affective psychosis. This may explain why manic-depressives with the right drugs live as entirely normal, socio-economically functional individuals more often than schizophrenics, for whom, with or without drugs to suppress the worst symptoms, integration is more difficult. Their problems are more intrinsic.

The dichotomy between these two types of individual difference in the brain—the transient, chemical imbalance and the permanent, structural one—is not, however, absolute. A long-term deficiency of a transmitter substance will cause the circuits using it to atrophy, so a chemical deficiency can become an anatomical one. On the other hand, anatomical damage will lead to low activity of the chemicals associated with the damaged pathways, so anatomical abnormalities are also chemical ones. Furthermore, the problems associated with anatomical abnormalities can still be treated by chemical means, for often the circuitry is present but partly impaired, and drug interventions can boost the activity of whatever circuitry is left in a compensatory way. Finally, within the past couple of years, the evidence has become incontrovertible that, contrary to long-standing belief, new brain neurons can grow during adulthood, in response to the right stimulation. The extent of this regeneration is not yet clear, but it implies that anatomical deficiencies may not be quite as irreversible as once thought.

◆

I have argued that there is no gene 'for' psychosis. Instead, there are many genes, the variants of which contribute to variation in the human personality, including along the dimension I have referred to as psychoticism. Individuals who are high up on this scale have an increased risk of developing psychotic disorders. With the brain science and genetics explained a little, I am now in a position to revisit a question that I touched on earlier in the chapter. This is the issue of whether the two psychoses are really distinct entities, or just alternative manifestations of the same process. This question can be rephrased from the dimensional perspective: is there a single psychoticism axis, or two different ones at right angles to each other?

The first possibility, that of a single dimension, was assumed in Eysenck's idea of a P axis. Eysenck believed in the fundamental unity of the psychoses, and thought that the differences in manifestation between them were due to other personality differences. For example, he considered the possibility, following a long tradition that ultimately goes back to Jung, that the manic-depressive and the schizophrenic differ only in that the former is an extravert and the latter an introvert. There is no difference between them in terms of the nature of their psychotic tendencies *per se*.

The second possibility requires us to identify two independent subdimensions relevant to psychosis, one relating to schizophrenia and the other relating to mood. The former already has a well-established name, schizotypy, whereas the latter lacks an accepted designation as yet. I might call it, by analogy, thymotypy, for present purposes. Being highly schizotypal predicts a liability to schizophrenia, whereas being highly thymotypal predicts a liability to affective psychosis.

The one- and two-dimensional approaches make rather different predictions. The one-dimensional approach suggests that there is a fundamental affinity between the two disorders; that the genes that make affective illness likely are also those that make schizophrenia likely; and that the brain imbalances are similar or related in both cases. The two-dimensional approach, by contrast, implies complete genetic and biochemical independence of the two disorders.

Some evidence can be found in support of both possibilities. The two disorders do seem, at first, to be very different, one starting from emotion and the other from cognition. The course and outcome are generally quite different. This supports the two-dimensional approach. However, as we have seen, patients can display a mixture of both types of symptom in varying proportions, and an intermediate diagnostic category, schizoaffective disorder, is needed. In terms of symptomology, then, there is no clear point of cleavage between the two. This supports the one-dimensional approach. A similar picture emerges from consideration of treatment strategies. The treatments usually used are different, though both antipsychotic drugs and mood stabilizers work on neurotransmitter systems. It is also true that many of the treatments associated with severe affective disorder, such as electroconvulsive therapy and lithium, can help some schizophrenic patients. As for brain abnormalities, the patterns of the two disorders are distinct, but then again, some of the characteristic irregularities associated with schizophrenia can be found in affective patients, too. The monoamine neurotransmitters, and enlargement of the cerebral ventricles, have been associated with both disorders. The evidence from medicine and neurobiology, then, is at best ambiguous in terms of deciding between the two approaches.

The crucial evidence comes from genetics. If the dimensions were really distinct, then there should be a tendency for psychosis to 'breed true'. That is, manic-depressives should have manic-depressive children, and schizophrenics, schizophrenic ones. If psychoticism is really unitary, then manic-depressives would be likely to have schizophrenic children, and vice versa. The evidence for breeding true is at best ambiguous. A well-known study by Professor Stassen and his colleagues in Zurich illustrates this. The study looked at the type of disorder found in the psychotic relatives of psychotic probands. The investigators categorized the disorders, in both relatives and probands, into four types: unipolar depressive, bipolar manic-depressive, schizoaffective (which has both types of symptom), and schizophrenic. The cross-tabulation of proband's disorder and relative's disorder is shown in Table 1. As you can see, there is a statistical preponderance of breeding true. Ninety per cent of the psychotic relatives of the affective patients had an affective disorder, and 48 per cent of the psychotic relatives of the schizophrenics had schizophrenia. However, there is also quite a lot of cross-over. Around 20 per cent of the psychotic relatives of affective patients developed schizophrenia, and over half of the relatives of schizophrenics developed an affective disorder. (These percentages do not add up to 100 because the diagnostic clusters used were not mutually exclusive.) The families of patients diagnosed schizoaffective patterned somewhere between the two groups in terms of the disorder profile.

Table 1 Cross-tabulation of proband's disorder and relative's disorder for a sample of psychotic patients.

Proband	Relative			
	Depressive	Bipolar	Schizoaffective	Schizophrenic
Depressive	73%	17%	2%	24%
Bipolar	46%	16%	4%	23%
Schizoaffective	49%	11%	0%	38%
Schizophrenic	43%	15%	2%	48%

From Stassen, H.H. *et al.* (1988) *European Archives of Psychiatry and Neurology*, 237, 115–123.
The rows do not add up to 100% because the diagnostic clusters are not mutually exclusive.

As well as this evidence from family studies, there is some evidence from cases of identical twins. Most often, psychotic twins will both have the same disorder, but there are a few cases where one is clearly schizophrenic, while the other is unambiguously diagnosed with an affective disorder. There thus seems to be a statistical tendency, but not an absolute principle, for the psychoses to breed true. There are several possible interpretations of this finding. One is that a one-dimensional model is correct, and that the breeding true that is observed is due to other factors, perhaps related to other aspects of personality, perhaps to gene–environment interactions. Another is that a two-dimensional model is correct, but that there has been assortative mating between schizoid and affective families, thus accounting for the cross-over that is observed.

This is an area of ongoing research, so for the rest of this book I shall discuss the matter in terms that attempt to embrace the positive truths in both one- and two-dimensional approaches. A two-dimensional model does seem appropriate. This is because breeding true is really the norm. Furthermore, as we shall see in Chapters 4 and 5, the brain disruptions in the two psychoses do seem to be of distinct, if overlapping, kinds. However, it must be conceded that the disorders are not entirely unrelated, as cross-over is just too frequent for that to be the case. It seems likely, then, that a two-dimensional model is basically right, but that the two dimensions are not completely independent. This means that high scores on schizotypy and high scores on thymotypy tend to go together more often than not, and therefore that a family high on the schizotypy scale would also be more than normally likely to be high on the thymotypy scale. This would account for the frequency of cross-over.

The best interpretation of this solution in genetic and neurobiological terms is that there are some genes that make individuals specifically more schizotypal or more thymotypal, and some others that load for both traits simultaneously. This explains how some of the brain and biochemical abnormalities in patients of both types can be similar, while others are different, and also explains the statistical association of the disorders. It is also still compatible with the finding that the psychoses have distinct courses and treatments, and usually breed true. Under this view, the dimension of psychoticism would be decomposable into two partly independent subdimensions, schizotypy and thymotypy.

Psychometric studies of psychoticism do indeed find personality similarities between affective psychotics and schizophrenics. They have also shown that the psychoticism dimension can be decomposed into several subdimensions. However, they do not, in general, find a simple binary split within psychoticism between schizotypy and thymotypy, the cognitive and affective parts. Instead, they find a more complex structure. The schizotypy/thymotypy framework is thus a simplification, but a simplification that will serve us well for present purposes.

Whatever exact formulation turns out to be right, the merits of the dimensional view are considerable, for it easily encompasses the cross-over between the two groups of psychoses, just as it naturally accommodates severity effects and the continuum between normal and psychotic functioning. We now have a fairly clear idea of the nature of psychosis, of the evidence for hereditary transmission, and of what is inherited. We must now turn to the disorders themselves, and ask where they come from.

CHAPTER 4

The storm-tossed soul

> So why should I want anything to do with [manic-depressive] illness? Depressed, I have crawled on my hands and knees in order to get across a room, and have done it for month after month. But, normal or manic, I have lived faster than most I know. And I think much of this is related to my illness—the intensity it gives to things and the perspective it forces on me. I think it has made me test the limits of my mind . . . always, there were those new corners . . . I cannot imagine becoming jaded to life, because I know of those limitless corners, with their limitless views.
>
> Kay Redfield Jamison, An Unquiet Mind

The year 1840 was a remarkable one for Robert Schumann. The previous 6 months had been unproductive and frustrating. Schumann was in Leipzig, while the love of his life, Clara, herself a brilliant musician, was stuck in Berlin, unable to come to him due to her father's objections to their marriage. This hit him very hard, and although he began working on many different projects, few or none of them came to anything. He was tortured by sleeplessness, lethargy, and a feeling that he was unable to compose any more. Only the prospect of Clara's arrival in Leipzig, which in fact was for no more than a court hearing in the continuing battle to have her father's objections set aside, was able to lift him from his torpor, and, during a brief burst of activity in the early winter, he wrote three romances for the piano.

After spending Christmas with Clara in Berlin, and with increasing reason to believe his marriage could be made to happen, Schumann felt his mood begin to lift, and in January he began to work in a more sustained way, starting with a sonatina for piano. On the 1 February, in ever better spirits, he turned his attention to songs, and thus began the 'Year of Song', one of the most extraordinary episodes in the history of human creativity. By December, Schumann had written over 130 songs, at an average rate of one every two and a half days. The peak rate must have been much higher, since the composer took time out for a blissful fortnight with Clara in Berlin in April, for Clara's arrival in Leipzig in June, and for their triumphant marriage in early September. The 20 songs of the Heine cycle were all completed between 24 May and 1 June; two and a half songs a day, by hand, which is about the rate a skilled copyist could make copies of them.

Periods of enormous productivity recurred throughout Schumann's life. In 1841, he sketched out a symphony in 4 days, and his productivity in 1849 actually exceeded that of the Year of Song, though by this time he had turned his attention to a wider variety of musical forms. Nor was his productivity restricted to the world of music; his life is littered with poems, plays, and essays, as well as the small matter of 2 operas, 19 choral works, 51 orchestral works, 320 songs, and 75 part-songs, and, depending on exactly how the counting is done, between 100 and 300 piano pieces. All of which were produced in just 46 years. When times were good, Schumann seems to have been capable of enormous, expansive joy at the world and his place in it. 'I am so fresh in soul and spirit that life gushes and bubbles around me in a thousand springs', he writes, and elsewhere, 'The entire heavens of my heart are hung full of hopes and presentiments. As proudly as the doge of Venice once married the sea, I now, for the first time, marry the wide world.'

Schumann's capacity for productive joy was not without its cost. Throughout his life, there were prolonged periods of terrible, anguished melancholy. In the Autumn of 1833, at the age of 23, he fell into deep despair following the death of his sister-in-law, and contemplated suicide by throwing himself from his upstairs window. He was unable to finish any of his compositions that winter, and, as so often later in his life, he became obsessed with the fear that he was going mad. His low spirits recurred, as we have seen, in 1839, and then, in more awful form, in 1844. That year, which was spent in Leipzig, in Dresden, and on tour in Russia, Schumann composed nothing at all. He was tortured by sleeplessness and fearful imaginings, and in the mornings Clara would find him swimming with tears. The most ordinary tasks became difficult, and music was intolerable, since he found it 'cut into my nerves as if with knives'.

Schumann did recover his spirits, to undergo another great period of productivity, but there were fewer than 10 good years left. In the winter of 1853–4, he relapsed once again. He became delirious, and was tortured by angels who sang beautiful music to him. He wrote down one theme, but, mostly, the music escaped him, and soon the angels turned to devils threatening to take him to hell, and then to tigers and hyenas, seizing him in their claws. On 27 February, Schumann rushed sobbing from his house and threw himself into the Rhine. He was rescued by fishermen, returned home, and was subsequently placed in an insane asylum, which he was never to leave. He died there, of self-starvation, in July 1856.

As well as Schumann's major depressions, there were countless smaller episodes of melancholy and anxiety, and frequent thoughts of suicide. There were also countless emotional highs; Clara was just the last of a series of adolescent infatuations, of both sexes, that he used to give himself a soaring romantic surge. Schumann craved these highs like an addict (he was also a drinker), and would structure his life, with a compulsive work schedule, to make sure he obtained them.

The times when the path to these rewards was blocked, such as when work was not going well, or when Clara was away or the attention was not on him, were the dangerous times, and these factors, or any negative shock from outside, could put him on the downward spiral. He was tortured by sleeplessness, nightmares, and waking dreams of almost hallucinatory intensity, often involving music.

In psychiatric terms, Robert Schumann (Fig. 6) was a manic-depressive, and he illustrates many features of the disease very clearly. The most obvious is the alternate cycle of high and low mood. When he was in his low moods, Schumann's world must have been intolerably bleak; at its worst it pulled him across into full-blown madness—hearing voices, believing he was persecuted by tigers and hyenas. Nevertheless, Schumann's depressive phase, though extreme, does bear some relationship to normal sadness. It was often triggered by the kinds of losses and frustrations that precipitate low mood in the most healthy of people: the death of a relative, or separation from the person he loved. It is just that the mood, once in, became an all-invading, resident, gnawing canker that stopped him doing the very things that would cheer him up.

Schumann's highs—such as that he experienced during the Year of Song—were probably not mania in the strict sense. This is because the clinical category of mania kicks in where the elation becomes entirely pathological. Schumman's highs were rather wonderful. Not only did he manage to work with nearly unprecedented effectiveness, but he also maintained normal life routines and a positive loving relationship with Clara. This kind of high would be called *hypomania*, an elevated but not yet psychotic state. If Schumann became so high that his thoughts were too disorganized to complete his work, or if he became frustrated and angry with those around him, or if he tripped over into grandiose delusions, then we would have to call it mania. If he had merely believed he was a great composer who had written hundreds of masterpieces, he would

Figure 6 Robert Schumann (reprinted with the permission of the Austrian National Library, Vienna).

have been deluded; in fact, he *was* a great composer who had written hundreds of masterpieces, so he was just hypomanic. Occasionally there are hints that his highs were so high as to be a hindrance, but in the same moment he captured the positive nature of his sweeping mental energy: 'But if you only knew how my mind is always working, and how my symphonies would have reached Opus 100, if I had but written them down ... sometimes I am so full of music, so overflowing with melody, that I find it simply impossible to write down anything'.

Once again, although Schumann's highs are clearly an aberration from normal behaviour, they have a clear precursor in the normal emotion of joy. Joy and hypomania are often sparked by the same things, romantic love being the most obvious example, and they have that same sense of optimism and energy to take on new challenges that fulfilled happiness can give us. There was also a seasonal pattern to Schumann's moods, with downs often coming in the autumn and winter, and his mood lifting in the spring and summer. Indeed, when his mood lifted, he often wrote of 'spring-like' feelings. This mirrors the more mundane glumness and gladness many of us feel in response to the changing seasons.

This chapter is about affective disorder, which is the cover term for the disorders of mood, depression, and mania. I have already given a taste, through Schumann's story, of what those disorders are like. I shall now be concerned a little with their biological basis, and a great deal with their relation to the normal experience of mood. This is because, the reader will recall, I have already argued that those with a liability to depression and mania are just those at the extreme of a continuum in the reactivity of their mood system. We thus need to examine the nature of that system in health, and consider why it is there. This turns out to provide many insights into the factors that precipitate depression and mania, and the strange behaviours that people in the iron grip of those conditions engage in.

◆

Both depression and mania are disturbances in the homeostasis of our moods. Moods are like emotions, but longer lasting; a brief sense of joy on seeing someone is an emotion, whereas if the happiness lasts for hours or days, it is a mood. Moods have both centrifugal and centripetal momentum. The centrifugal forces arise from the fact that moods can be self-reinforcing. A mood of sadness makes us more prone to remember other sad things, and it makes us withdraw from normally pleasurable activities, thus depriving us of the very things that would lift the mood again. By contrast, happiness makes us feel optimistic and self-confident, so we take on rewarding tasks, behave in an attractively outgoing way, and tend to remember positive things. All of these give us more reason to be happy.

The centripetal forces come from the fact that moods can also be self-limiting. Something makes us sad, so we stop doing it, change our situation, or address the source of the sadness by finding a better way of thinking about it. Gradually, the sadness feels less bleak. Alternatively, when things make us joyful, we tend to continue them, but may feel our joy diminish as we tire, the novelty fades, and we get less out of them.

In normal functioning, the centripetal forces generally outweigh the centrifugal ones. Our mood moves up or down in proportion to the things that happen to us, but it has a natural tendency to return to centre, as a result of our own thoughts and actions as well as the course of time. This is emotional homeostasis.

In depression and mania, the balance of forces is reversed, with the equilibrium not in the calm centre but frighteningly far out towards the edges. A negative event, like the death of Schumann's sister-in-law, has an understandably saddening effect. However, the depressive, instead of a slow but buoyant return to normal, spirals the other way. The extreme sadness about one thing brings other negative thoughts and worries flooding in; these make him even more sad, which continues the negative thoughts, and so on round the loop. His subdued lethargy, and pessimism about what he can achieve, make him take on less in terms of professional challenges; this deprives him of the highs of success, and gives him good cause for his subdued pessimism, and so on round the loop. He feels he is unattractive socially, so he becomes dull and withdrawn from interaction, which means he *is* unattractive socially, which he perceives, and so on round the loop. He does not sleep, so he is in poor form, and does hurtful things which make him anxious, and the anxiety keeps him up at night, and so on round the loop.

In mania, the spiral goes the other way. The sufferer is on such a high that he takes on new projects and plans. He may start new sexual liaisons, which he experiences with unbelievable passion, and has the optimism required for numerous creative endeavours and financial risks. These give him such a buzz that his optimism and energy are renewed, nay, pushed even higher, and so he takes on more new things, and so on round the loop. This would be all very well, but he soon becomes exhausting or manipulative in his quest for experiences, and when the world does not respond to his racing fancies, he can become angrily frustrated with it. At the extreme, his mind races too much, he does not separate what he fantasizes he has done from what he actually has done, and he may lash out at the dull mortals who impede him. He may be financially and emotionally ruined; he may be quite mad.

Both depressive and manic episodes are usually self-limiting in the end. That is to say, eventually, the intrinsic centripetal forces of the mood system will counteract the centrifugal spiral, and bring the person back to centre. However, this is no consolation whatever to those in the midst of the storm. In severe cases the sufferer will be dead before self-limiting has had a chance to happen, and, much more often, he will have lost the vital things that gave him strength in his life: partner, job, friends. This, of course, gives him good reason for staying depressed, or for plunging into depression if his episode was a manic one. For these people, to quote the manic-depressive poet Byron, 'Their breath is agitation, and their life/A storm whereon they ride'.

I have thus described the psychology of mood as one of centripetal and centrifugal forces, and of mood disorders as out-of-control centrifugal spirals. Interestingly, the biology of mood can be thought of in much the same way. A reasonable account of this biology can now be given, thanks to a large body of research carried out on people and, where this is impossible, on monkeys, who do not seem to differ much from us in brain structure and chemistry at the emotional level, however much we may have departed from their line in terms of cognition.

Let me start the account at the point where a person confronts a challenging stimulus. This has many effects. The systems of the brain that detect the emotional connotation of a stimulus, which include a structure called the amygdala and parts of the prefrontal cortex, fire up. These systems are fed by neurons that use monoamine neurotransmitters, and so there is a surge in monoamine transmission. The systems that make us seek reward seem particularly associated with dopamine, while those that defend us from harm (by making us anxious and vigilant) seem especially linked with norepinephrine. Serotonin systems function to make us adapt to psychological challenges; on repeated application of a stressful stimulus, experimental animals show a surge of serotonergic transmission. This is part of getting used to a new challenge, and no longer feeling anxious about it. The serotonin systems probably achieve this, at least in part, by their regulatory effect on the systems of the other monoamines.

The response to a challenging stimulus is not restricted to the brain. The brain communicates with the body not just with direct signals to muscles, but also, via the pituitary gland that hangs at its base, with changes to the levels of hormones in the bloodstream. Hormones are as important as nerves as a means of communication and command in the body, but the way they work is different. Whereas nerve signals to muscles are short-term and highly specific messages, hormonal changes are longer acting and rather generalized. Nerve signals make us move our left elbow, whereas hormonal changes make us more awake, or more asleep, more hungry, or lustful, or more or less sensitive to pain. In particular, the set of related blood chemicals known as the stress hormones, which include cortisol, gear us up for big physiological challenges. They increase the heart rate, make us alert, and divert blood flow from the digestive system to the muscles.

The stress hormones are also implicated in disorders of

mood. Their level is elevated in many depressed people, as well as in those undergoing major psychological adversity. They seem to be involved in a feedback loop with serotonin systems. Serotonergic circuits are stimulated in response to sustained adversity, and probably help damp down the stress response. On the other hand, it is clear that high levels of stress hormones impair the functioning of serotonergic circuits. The control link between serotonin systems and stress hormones is thus a two-way one.

This brief foray into physiology allows us to see in outline how the centripetal and centrifugal spirals must work. In normal functioning, the challenging stimulus (which can be either negative, leading the person to avoid harm, or positive, leading the person to seek reward) leads to increased arousal in the relevant parts of the brain, governed by monoamine transmitters, and a stress response in the body, through the hormonal cascade down from the pituitary gland. As the person copes, by avoiding the stress or achieving the reward, the response from those systems subsides, and, more generally, serotonergic systems activate to damp down the anxious responses.

In disordered mood, this homeostasis is not achieved. The challenging stimulus may cause the anxious arousal and stress response, but, rather than coping, the person is sent on a downward spiral. Very high levels of stress hormones impair serotonergic functioning, and with serotonin impaired, there is nothing to damp down the further cascade of stress hormones or regulate the function of the other monoamines. Thus the state of anxiety and unhappiness is self-perpetuating. The biochemical imbalance also affects a brain structure called the hippocampus, which is implicated in our memory for events, and makes depressing or anxious memories more likely to come to the fore. These, in turn, cause further anxiety and stress responses, and so on. The levels of neuroreceptors attempt to adjust to the changed levels of neurotransmitters, but they do not seem to compensate adequately, and the depressed mood takes up residence.

In mania the operation of the spiral is the other way, up rather than down. However, the mania spiral is not exactly the mirror image of the depressive one. The evidence suggests that serotonergic function is reduced in the depressive spiral, but rather than being increased, it is also reduced in the manic one. Remember that the serotonin system is the regulator of the other monoamine systems, and thus mania and depression may also be alike in starting from a serotonergic deficit which permits the other systems to go out of balance. The difference between them is what happens in the other, now-unregulated circuits. There is some evidence that, in mania, norepinephrine levels are elevated (which may account for the grandiosity and high level of activity) and that dopamine levels can also become elevated (as they are in acute schizophrenia, which may account for the schizophrenia-like symptoms of delusion and hallucination, and for the frenetic search for new experiences).

This biological account gives us some insight into how a person's underlying disposition and life events must interact to produce a depressive episode. Unresolved life problems cause stress, which elevates the level of stress hormones such as cortisol. These eventually impair the balance of the serotonin system, and perhaps, by consequence, other monoamine systems as well. Individuals who are constitutionally vulnerable to affective disorder carry genetic variants coding for monoamines and related chemicals, which means that their system is easily disturbed. For them, therefore, the severity of stress that is required to cause dysfunction is not too high. Individuals low in vulnerability, by contrast, would need a much greater stress input to cause a prolonged mood disturbance. It could happen, but it would take far greater life stress to do it. We also get some idea how nurture can play a role by the fact that distressing experiences could either permanently atrophy the serotonin-using brain systems, or set up stressful psychological associations, which mean that later life problems provoke stronger stress responses, thus eliciting the downward spiral more easily. It is well known that chronic stress, and abandonment or the loss of a parent early in life, do increase the risk of later depression, although the contribution of these factors to its incidence is thought to be much less powerful than the genetic contribution.

Note, too, that the kinds of initial stimuli that start the downward spirals can be very varied. It does not matter if the initial challenge is a difficult life event, a drug taken for some other condition, or a minor brain injury such as a stroke. From the brain's point of view, these all look much the same; they all equate to changed flows of brain chemicals and hormones, and therefore any of them can trigger an episode of mood disorder. Similarly, anything that can interrupt the spiral can serve as an antidepressant. This can be an affirming life experience; a therapist who can change beliefs and emotional associations; a biochemical spanner, such as Prozac, in the spiral's works; or even an electric shock to the brain itself (as administered in electroconvulsive therapy, which is less dangerous than its reputation insists and is still a highly effective short-term intervention for really severe depression). There is no philosophical problem about how such different therapies as these can all produce the same effects. It is only from our everyday perspective that they seem different in kind. For the brain, they all get cashed out in the same biochemical fashion.

It is clear from this little glimpse of how the monoamine and stress hormone systems work that we humans are equipped with a complex, finely tuned system for regulating mood, a system that fails in affective disorder. But the existence of such a system naturally raises a further question, what is it there for? Why does the brain have a mood system at all? To address questions of this nature, we are always forced back to thinking in evolutionary terms, for only evolutionary reasoning can explain why the brain has the design features it does. The four-chambered design of the mammalian heart is related to its function in pumping blood. That is to say, those mammals with hearts of this type ultimately outreproduced all their competitors. A similar principle must be applied to the primate mood system. Those among our ancestors who had it must have done better than those who did not, and in this respect it must relate positively to the brain's general functioning.

What are brains for? Their chief function seems to be to produce behaviour that is appropriate to its context; and, in evolutionary terms, appropriate behaviour is just whatever makes the organism more successful at reproducing. The term biologists use for traits that do this is *adaptive*; a trait is adaptive if it enhances the reproductive success of the animals bearing it, and I shall use the term in that sense here. Note that reproductive success is not the same thing at all as number of offspring. An animal that had thousands of offspring, none of whom produced offspring themselves, would have lower reproductive success in the end than an animal with one son and one daughter, both of whom also produced one son and one daughter. Maximizing reproductive success is about maximizing your long-term representation in the gene pool. One way of doing this is to spawn like a fish, with many thousands of offspring, some of whom survive, with a small investment in each one, but this is not the only strategy. In other circumstances, it can be adaptive to have a small number of progeny, and invest your resources massively in each one. Late-twentieth-century humans are perhaps the best example of this approach.

In order to understand how having a mood system is adaptive, we have to understand some common features of the primate world, features that apply to all our fellow primates, and even more strongly to ourselves. The first is that primates are intensely social animals. Most species of primate, particularly the monkeys and apes, live in large, dynamic social groups. These groups, although variable from species to species, have two common attributes. Firstly, there are hierarchies of status, and an animal's place on these status hierarchies is a strong determinant of its reproductive success. More dominant males have access to more females; more dominant females have access to better protection for their young; and dominant animals of both sexes have better access to the 'economic stuff' (which is food, for all species other than us, and something rather a lot more complex in our case).

The second important attribute of primate groups is that, within them, individual social relationships are a source of power. All primates have, in rudimentary form, that central characteristic of human life—we get on with some people better than others, and we look out for our friends, often planning with them to mutually enhance our position. In primate groups, friends and allies support each other, on a tit-for-tat basis, in bids for status and resources, and so an individual's social bonds are of great importance in determining his reproductive success.

Primates, then, are social. They are also very brainy. These two features may not be unconnected. Indeed, the most promising current theory for why primates are so brainy is that the calculations and machinations of living in a complex social group demand more cognitive power than is ever needed by an animal living alone. The evidence for this theory is simply that the more social a primate species is, the larger, in relative terms, is its cerebral cortex, and in both of these respects *Homo sapiens* sits at the top of the pile. The consequence of primate braininess, which is relevant to my argument, is the kind of computational power it provides. Because of their enlarged cortex, primates, and especially humans (who take this into a new league) are very good at reasoning in multiple steps from cause to effect. They excel at the kind of computation that goes: 'if I do this to him while pretending to be her, he will believe that she wasn't with me, and he will stay away'; or, to transfer the same thing away from the social domain, 'if I fill this hole with this plug, this bucket will fill up and become heavier, and the branch will break'. Sophisticated reasoning of this general type is the very basis of all the human achievements you can name; Restoration comedies and soap operas, in the former example, chess and horrendously complex train sets, in the latter.

So, we have to achieve our adaptive goals in a complex social world, and we have a powerful computer to help us do this. But even the most powerful computer is lost unless it is told what the objective of its computations is, and here is where the emotions, and the mood system in particular, come in. It seems that emotions serve to direct us to the kinds of things that are likely to be adaptive. Pain makes us withdraw from harmful things, whereas fear makes us avoid things likely to damage our reproductive success. Of course, the genes cannot know exactly what we are going to encounter in life, so cannot equip us with specific behavioural programmes for dealing with all eventualities. What they can do instead is give us a powerful computer for working out how to get from A to B, and a few very nice or very nasty feelings to help us work out which of A and B we should be heading for anyway. If A tends to damage our prospects, we become fearful, angry, or hurt towards it. If B tends to do our prospects good, we have a range of emotional connotations, such as liking and loving, which we find rewarding and which will draw us back to it again.

The mood system plays its part in all of this by modulating our aspirations in the world. Remember the behavioural characteristics of high and low mood. In high mood, the person stands proud, feels self-confident, reports optimism about the future, and tends to rate his abilities highly when asked. He is more likely to take on new personal relationships or professional plans, and more prepared to take risks to enhance his status. This becomes reckless in mania, but in the milder form it is the basis of much positive achievement. In other words, in the high-mood individual, the cognitive computer is set to work with a general goal of moving up hierarchies of social success and status.

In low mood, the reverse is true. The low-mood person is

pessimistic about his own future. He rates his own abilities and aptitudes negatively, and is reluctant to take on new projects. He withdraws from new relationships and concentrates on the safety of the known. His very posture is that of submission; the rounded shoulders and the lowered gaze make him look less than he is. The computer works with the goal of minimizing risks, of consolidating the known, and of saving energy. In the extreme of low mood, depression, it is common for people to feel hopelessly unworthy of their partners or their jobs; their self-perceived status has dropped, and aspirations are set lower than before. Unfortunately, there is nothing so unattractive as this kind of self-perception, and the feelings often become self-fulfilling.

How can low mood be adaptive, given that in all primate groups, status is positively related to reproductive success, and low mood makes us drop in status? There are two answers to this, because moods have two kinds of function, which we may call *aversive* and *adjustive*. The aversive function stems from the fact that low mood is intrinsically unpleasant. It is unpleasant enough that, if we find out that a certain action brings it on, we will tend to avoid that action in the future. Thus, by attaching a nasty tag to things that caused a decline in our adaptive status in the past, it helps steer our cognitive computer clear of them in the future. High mood, of course, has the opposite effect. It is intrinsically pleasant, and helps guide us back to those roles and behaviours that improved our prospects in the past.

The adjustive function of low mood helps us find the right level for ourselves. Although it is better, in evolutionary terms, to have high status than low, the worst thing of all is to try to maintain a status that is not actually justified. Animals that attempt to take the food or reproductive privileges that go with high status without the strength and social back-up to do it have a terrible vengeance exacted on them by more dominant animals. They can be killed or driven out of the group. What mood does is help accord an individual's plans with what he can actually pull off. This explains why the common triggers of mood are what they are.

Low mood is brought on by the loss or frustration of those things that would enable one to move higher in social hierarchies, and enhance one's adaptive goals. The most obvious ones are the death of, or falling out with, close friends, relatives, or parents; failure to achieve some professional goal that would have brought status and resources; and, of course, the loss of a worthy partner. (A worthy partner is, after all, the high road to Darwinian heaven, so it is no accident that bereavement increases the risk of major depression by about sevenfold, or that divorce is also a major trigger.) All of these things represent a lowering of social position and a loss of opportunities for reproductive success. Following them, the mood drops, and the person's goals are reset to a more appropriate level.

High mood has the opposite adjustive effect. A professional or artistic success makes us more optimistic about the future, and makes us seek out new relationships or partners. Or it can be the other way around; a personal success makes us braver and more energetic in some professional domain. Either way, there is a positive upgrading of adaptive goals. Success in one domain makes us place our estimate of position higher, and gives us confidence to seek success in other domains as well.

I have argued, then, that the mood system (in normal functioning) is adaptive because it helps us set our life goals in a way that is appropriate to our experience and position. It does this by tagging certain behaviours with negative or positive connotations, and allowing us to adjust gracefully to changes in reproductive potential. This also explains a great deal about the seasonal patterns of mood that occur. The autumn in the temperate latitudes is associated with a peak of suicides and depressive symptoms (known as seasonal affective disorder), and even in non-depressives, there may be more subdued feelings. This makes sense when you consider that, in the past, declining temperature and day length were usually associated with declining resource possibilities, so it was as well to adjust one's plans downwards. This is why it is hard to sell a house in England in the winter; quite aside from practical considerations, people are less willing to take on that kind of big risk. In the spring, on the other hand, improving conditions lead many people to feel 'mad as March hares', and as Chaucer observes at the beginning of the *Canterbury Tales*, once April is in, people look to take on adventures. Clinical mania also tends to be more common in the summer months. (The picture is not quite as simple as this; there is also a peak of depressive symptoms and suicides in the spring. This is probably because, for people with imbalanced mood systems, any change in mood triggers risks sending them on a downward spiral.)

What is the evidence that the mood system has the functions I have ascribed to it? For one thing, it makes sense of our observations on the causes and consequences of human moods. It also makes sense of the characteristic behaviours associated with disordered mood. The grandiosity and financial recklessness of mania are overoptimistic assessments of social potential, whereas the suicide commonly associated with the most severe depression is also interpretable from this standpoint. After all, what are suicidal feelings, but the most extreme negative assessment possible of the future potential of living?

As well as this indirect evidence, there is a series of direct studies of the mood system in vervet monkeys by Michael Raleigh, Michael McGuire, and their colleagues. The social groups of vervets, like those of other primate species, have clear hierarchies of status. In these groups, the dominant males have serotonin levels one and a half to twice those of their subordinates (you cannot measure brain serotonin levels directly, remember, so these levels are measured in the blood). The dominant animals also exhibit the classic positive behaviours stimulated by serotonergic treatments such as Prozac in humans—they are less aggressive and anxiously vigilant, and they engage in affiliative behaviours, such as grooming, more often than subordinates.

If the dominant monkey is removed from a group, a competition for dominant status ensues, and, interestingly from our point of view, the male that wins then undergoes a rise in serotonin levels, and his behaviour changes accordingly. This happens whichever animal it is that wins—the achievement of status lifts the mood. Life is not so sweet for the dominant monkey that has been removed from the group. Kept apart, his serotonin levels drop to those of a subordinate individual. Presumably for him, it is as if he had lost his status and been cast out, and his low mood reflects this.

Competitions for dominance can be manipulated by intervention with drugs. If, when the dominant monkey is removed, one randomly selected male is given a serotonergic enhancer such as Prozac, that male becomes the dominant one *in every case*. He seems to achieve this by frequent and confident affiliation with females, who rally around him. Conversely, monkeys treated with serotonin-depleting agents during a dominance competition become withdrawn and aggressive, and never achieve dominance. These studies are interesting not just because they provide direct evidence of the relationship between social aspiration and the mood system in primates, but also because they reveal the self-fuelling nature of moods. The only difference between the Prozac-treated monkeys who achieved dominance and their competitors who did not was that the treated ones had been given an artificial 'lift' to the monkey equivalent of their self-belief. This parallels an equivalent finding in depressed humans. If you think you can do it, you often can do it. Conversely, if you think you can't, you often can't. Because of the evident power of self-belief, one of the most influential approaches to treating depression, cognitive–behavioural therapy, attacks the beliefs that patients hold about themselves as the major proximate cause of the problem.

In human societies, there are no simple hierarchies of dominance and subordination, nor any single path, up and down which our aspirations wander. Our paths to self-fulfilment, unlike those of the vervet (which seem pretty straightforward), are variable, nebulous, and culture-specific. Nor is there any evidence for the existence of mood disorder in monkey populations. None the less, we can intuitively see the relevance of the monkey studies to humans, for their simple social world is a kind of archetype for our own.

◆

I have given you a synoptic view of the world of affective disorder. It is a simplified view, too. Affective disorder, in fact, takes many forms, which are related to, but slightly different from, the classical versions of mania and depression as I have presented them here.

For one thing, there are great variations in the severity of affective problems. Most depression is relatively minor, and certainly does not qualify as psychosis of any kind. However, it is my belief, elaborated in earlier chapters, that the extreme, psychotic forms, and the minor ones all exist on a continuum. This continuum runs from occasional minor depressions at one end to full-blown bipolar psychosis at the other, through the intermediate disorder of major unipolar depression.

The social patterns of major and minor depression are somewhat different. Minor depressive problems are most concentrated amongst the poor, the unemployed, those in poor housing, and those with little social support. The hereditary factor may be less strong than for the major forms of the disorder. These minor forms are especially amenable to treatment with drugs such as Prozac, which elevate serotonin levels.

The major forms, on the other hand, are more familial, more recurrent, and have a different social distribution. Bipolar affective psychosis is actually most common in the professional classes and amongst the relatively affluent. These differences have led some authors to posit a fundamental difference in origin between the minor and major forms of affective disorder. According to this view, minor depression is just low levels of serotonin, brought about by low social status and lack of pleasure-producing conditions. The major conditions, on the other hand, are caused by an inherited impairment in the ability to regulate serotonin metabolism. Thus they are more familial and less environmental in their causation.

This dichotomy is simplistic. The minor forms of affective disorder are very common in the families of those with the major forms, so the two are obviously genetically related. Furthermore, there is no non-arbitrary dividing line between major and minor. Fortunately, the dichotomy is also unnecessary, in view of the dimensional perspective on psychiatric disorder that I have argued for in this book. Instead of two different kinds of affective disorder, I would argue that there is a continuum of predisposition, related to the continuum of efficiency of people's serotonin systems, which interacts with social conditions. Poor social conditions can produce depressive symptoms even in people with a relatively low level of predisposition. However, if the level of predisposition is low, the symptoms will be relatively mild. Hence the concentration of minor, relatively non-familial, depression in lower social classes.

A high level of predisposition, on the other hand, will shine through, whatever the social conditions. This accounts for the occurrence of bipolar affective psychosis in many successful people who have apparently charmed careers. It also accounts for the more obvious familial pattern of the major disorders. (The question of why those families with a predisposition to affective psychosis should so often be of high social standing is one to which I return in a later chapter.) Between the two extremes of entirely socially generated minor depression, and entirely familial bipolar affective psychosis, there are, of course, many intermediate cases.

As well as variations in the severity of affective disorder, there are variations in the form it takes. Affective disorder has been linked, genetically and biochemically, to many other conditions, such as generalized anxiety, alcoholism and other substance abuse, eating disorders, criminality, and impulsive violence. What links all these apparently disparate conditions to depression is, when you think about it, an excess of negative feeling. In classical depression, the sufferer is rather overwhelmed by negative feeling, and feels passive in the face of it. In the other forms, the sufferer reacts in different ways. In alcoholism or substance addiction, he seeks to administer himself a pleasure fix to fight the negative feelings. In violence and criminality, he lashes out against the social situation he may feel is responsible for his feelings of lowness. In eating disorders, she adopts the deluded belief that the form of her body is the source of the problem, and punishes it with excessive control of eating. In as much as serotonin is the master key to the balance of well-being, these are all serotonergic problems. We can even add mania to the same list. Although mania, with its great elation, appears to be the opposite of depression, the frenetic hyperactivity and constant pursuit of novel pleasure can be seen as a desperate attempt by the sufferer's brain to compensate for a basic lack of calm well-being. Thus mania, too, may join the list of alternative responses to a common serotonergic deficit.

The variety of forms of affective disorder links in to another aspect of its distribution, that of gender. All studies agree that classical depression occurs about twice as frequently in women as in men in the Western population. Interestingly, there is no gender imbalance at all in the occurrence of bipolar affective psychosis, which is the most genetic and most serious of all affective disorders. This suggests that the underlying distribution of liability is the same in men and women. When the liability is really serious, as in bipolar sufferers, the disorder surfaces at the same rate. The excess of more minor cases in women might be explained by the existence of worse life stressors for women in our societies (these include, particularly, lack of support during child-rearing, the juggling of home and work life, and childbirth itself, which causes such hormonal fluctuations that it puts a woman at worse risk of depression than at any other time in her life).

Alternatively, the excess of classical depression in women might be because the men are responding to their problems of low mood in different ways. This is suggested by the fact that, although depression is less common in Western men than in women, alcoholism, criminality, and violence are vastly more common. Perhaps the men are responding with these, more outward-directed strategies, whereas the women are internalizing the problems much more.

There is evidence to suggest that this interpretation is the right one. For one thing, when the numbers of men who are alcoholic or impulsively violent are added to the numbers who are depressed, the total is about the same as the number of women who are depressed. More intriguing evidence comes from the Old Order Amish, the fundamentalist Protestant sect who live a largely traditional lifestyle in isolated communities in the USA. Within these communities, substance abuse is non-existent, violence is shunned, and the crime rates are virtually zero. Thus the alternative avenues for men with affective problems are simply not available. The result is that the men get depressed just as commonly as the women. Studies from other traditional societies suggest the same pattern.

These, then, are the affective disorders. They are amongst the leading killers and sources of suffering in the developed world, and the experience of a major attack is a bewildering, betraying, and profoundly inexplicable one for both the sufferer and his family. Such comments apply equally well, if not better, to the other category of psychosis, schizophrenia, to which we now turn.

CHAPTER 5

The sleep of reason produces monsters

> Should I let anyone know that there are moments in the schizophrenia that are 'special'? Where there's a different sort of vision allowed me? It's as if . . . I've gone around the corner of humanity to witness another world where my seeing, hearing, and touching are intensified, and everything is a wonder . . . I won't tell myself it's all craziness.
>
> M. E. McGrath

Adolf Wölfli was born in Switzerland in 1864. After a traumatic childhood, he spent his young adult life in a disjointed series of different residences and manual jobs. He was a loner who was unable to commit to anything, and he was eventually arrested for a sexual assault on a minor. In addition to his antisocial behaviour and indifference to others, Wölfli suffered strange hallucinations and delusions of persecution. He was frankly psychotic, and so, rather than staying in prison, he was sent to the Waldau mental asylum near Berne, where he was to spend the last 35 years of his life.

For his first 4 years at Waldau, Wölfli was too agitated to apply himself to anything, but as time passed he became more settled, and absorbed his energies in drawing. The drawings were of a bold, naive style, and executed with simple materials and remarkable skill. These early drawings foreshadowed the extraordinary artistic, literary, and musical output that was to occupy Wölfli for the rest of his life.

Wölfli went on to draw and paint on a much more ambitious scale, for which he is best known. However, in addition to his scores of pictures, he produced a remarkable literary and musical oeuvre. He has left a sequence of manuscript books that run to 25 000 densely filled pages. The full scope of this creation is only just becoming understood, but it is clear that it is a remarkable multimedia endeavour that was far ahead of its time. Wölfli was obsessed with order, and the organization of his vast work reflects this. The books are divided into five major series, each of which took several years to create, and each of which has a different structure. The first series, *From the Cradle to the Grave*, is an autobiography, factual in places, fictional in others, in which Wölfli assumes the role of an explorer and travels around Europe, often accompanied by imaginary siblings and companions. In the second series, *Geographic and Algebraic Books*, the narrative becomes more overtly mythological. Adolf Wölfli becomes the mythic hero St Adolf, and the places visited start to be magical islands and kingdoms, peopled with queens, gods, strange fauna, and bizarre forms. In the later books, the mythical places and characters continue to be explored, but the narrative structure becomes more diffuse. Instead of a sequential storyline, the books are organized around fragments of music and impressionistic tone poems, all written in the same hand in continuous sequence, which the interspersed prose explores and illustrates. The later drawings and paintings illustrate the people and scenes from St Adolf's mythical journey.

The originality of Wölfli's art was soon apparent. The pictures eschew naturalism for a strange, geometrical, almost mechanical style. Naive faces and animals are juxtaposed with abstract patterns, with prose, and with musical notes. By the end of his life, Wölfli was selling pictures fairly successfully, and posthumously he became one of the central figures in the 'art brut' or 'outsider art' movement. This movement, and Wölfli in particular, attracted interest from the surrealists, and in many respects he is a true avant-garde figure; the diffusion of a bold, disjointed, non-naturalistic style such as his is

THE SLEEP OF REASON PRODUCES MONSTERS 115

Figure 7 Hautania and Haaveriana, 1916, by Adolf Wölfli (reprinted with the permission of the Adolf-Wölfli-Stiftung, Bern).

detectable right across the twentieth-century arts, from surrealism and pop art on the visual side, to their equivalents like dadaism in poetry, the theatre of the absurd, and atonality in music. Wölfli's pictures have been exhibited widely in Europe and the United States, and are permanently displayed in his native Switzerland.

The literary work is less well known, and while no-one claims that it is easily readable or artistically successful, it too is clearly highly original. The intricate organization of the entire structure contrasts with a sequence of extremely opaque leaps of imagination in the flowing prose. Indeed, language itself must be distorted to follow this sequence, as the author coins a constant stream of neologisms and puns to follow his stream of thought. The effect is surreal, and brings to mind later postmodern novels and poetry. Above all there is a sense, which the writing shares with the pictures, of a profound dislocation of perspective in a world where nothing has its usual scale, form, or significance. A short extract from the *Geographic and Algebraic Books* suffices to show this imaginative strangeness:

> There's God-Father-Thunderbolt-Springing-Fountain, with a diameter 9 hours long, standing 1,999,900 hours high, in the Giant-City inhabited by 135,000,000 souls and a God-Father-Slaughter on the identically named Giant-Island in the large East-Ocean! And the Garden of Eden! And the Cedars-of-Eden, 1,999,900 hours high . . . We just arrived, the whole organisation of the Swiss Hunters and Nature Explorers Avant Garde, a total of 50–60 persons with their Four-Suspension-Axles-Iron-Bridge-And-Constant-Vehicles to the above-mentioned island. Already on its North Shore we heard a dull portentous thundering noise, and, after following the fine and comfortable route of the First-Rank-Street in the midst of opulent and majestic Southern cultures, flora and vegetables, we finally reached the city God-Father-Slaughter-North, and, before our eyes, we saw nothing except water, and finally still more water. And, to be exact, up to the highest altitude.

The category of schizophrenia did not exist when Wölfli was hospitalized (although its intellectual precursor, *dementia praecox*, was just being created by Emil Kraepelin). However, this is almost certainly the disorder he was suffering from. The striking oddness of his world brings to mind the inner life of the other schizophrenic we have met in this book, Mr Matthews (see Introduction). There are some obvious differences between them. It seems that Wölfli recognized the distinction between his literary creations and reality. Indeed, artistic creation may have been his way of dealing with the flood of strange experiences he had. For Mr Matthews, in contrast, the world of his fantasies was totally real. The content of the two men's imaginative concerns was entirely different, in fact unique; yet somehow there is a commonality in their inner lives, to do with profoundly abnormal sequences of perception, reasoning, and motivation. This is the conundrum of schizophrenia, that oxymoronic category, the unifying feature of which is idiosyncracy.

♦

One of the few statements about schizophrenia which is universally agreed on is that it is still the subject of much disagreement. The disorder has been in our conceptual universe for nearly a century, initially as Kraepelin's *dementia praecox* until Eugene Bleuler renamed it schizophrenia in 1911, but time alone has not been enough to settle all the battles that have raged around it. Indeed, it is still sometimes referred to as the central mystery of psychiatry.

A few facts are clear enough. There is no doubt that what is commonly called schizophrenia is a very serious social problem, affecting at least 250 000 people at any one time in the United Kingdom, and at least 1.25 million in the USA. In 1990, the annual costs of treating it in those two countries were estimated at £2–3 billion and $15–20 billion, respectively, and this is merely the tip of the iceberg, since the disorder has far-ranging indirect social and economic consequences. Between one-third and one-half of homeless people in the USA are thought to be schizophrenic.

Beyond this point, consensus breaks down. At one end of the debate, there are the extreme positions of those critical of psychiatry, such as Thomas Szasz, whose arguments we met in Chapter 1. They contend that schizophrenia is not a medical category but an ideological one, a label society puts on those people whom it finds too different to tolerate. Schizophrenia is certainly about difference in beliefs and behaviour, as we shall see, but, as I have already argued, this does not mean that a medical perspective is inappropriate. Extreme deconstructionist positions, such as that of Szasz, were popular in the 1960s and 1970s, but have lost much of their sting as more effective psychiatric drugs have been discovered, and evidence for the involvement of genes has quietly amassed. And amassed it has; as we saw in Chapter 2, the evidence for genetic factors in schizophrenia has proved much firmer than that supporting social causes alone.

Further infield, but still critical of the schizophrenia concept, lie those psychologists and psychiatrists who accept the medical perspective, but challenge the unity of the syndrome. They point out that the diagnosis of schizophrenia has been applied to patients with a wide range of disturbances in thought and feeling. The mix of symptoms, not to mention the content of delusions, varies from person to person. Schizophrenics have not all been shown to share an underlying brain pathology. Nor is the course of their disease uniform; some people have a single episode, with a sudden onset and an apparently total recovery, while others have an insidious onset and a degenerating history with no remission. Finally, people's response to treatment is quite variable. There are antipsychotic drugs that suppress many of the most obvious symptoms in most cases, but their long-term efficacy and specificity as cures are still a matter of debate. They often have a sedative, depressing effect that may actually exacerbate negative symptoms. In other cases, lithium, antidepressants, or electroconvulsive therapy may be employed. Some cases are severely resistant to all treatment. The evidence for a single disease entity is thus weak.

A reasonable response to these charges can be mounted. The diagnostic category of schizophrenia has been greatly sharpened up in the past 20 years, and while it remains true that the symptomology is very variable, some core phenomena can be identified. As for the biological mechanisms involved, mounting evidence, which I will review shortly, points to subtle brain abnormalities. No single abnormality has been found that is universally and uniquely diagnostic, but there are reasons for this, which are not just to do with the difficulty of research in this area, but also with the very nature of the condition, as we shall see. The variability of treatment response and prognosis are not so strange, in fact many systemic diseases show such inconsistency. This is especially true of psychiatric disease, where the personality and life-course of the individual are so central to the outcome. Furthermore, there have been many attempts to split schizophrenia into several subtypes, along the lines of symptom clusters, course of illness, or treatment response; these have not proved predictively valid. The subtypes do not divide up neatly, and certainly do not breed true.

The mainstream of psychiatric opinion, then, endorses the following view. Schizophrenia is a heterogeneous but none the less valid category of severe mental illness. The core features of the illness can be divided into three main types. First, there are the so-called positive symptoms. These are the bizarre beliefs, delusions, and hallucinations which are most often associated with schizophrenia in the popular imagination. Mr Matthews' strange beliefs, described in the introduction to this book, are a classic example. Typical features of positive symptoms include hearing alien voices in one's head (in Mr Matthew's case, the gang of villains), feeling that one's thoughts or speech are being controlled or manipulated by an outside force (for example, the CIA, or an 'influencing machine' like Matthews' air-loom), or believing that neutral, outside events have a special significance referring to oneself (these are called ideas of reference).

In contrast to the positive symptoms are the negative symptoms of schizophrenia. These are considered negative because they involve a diminution of emotion and motivation. The schizophrenic is often withdrawn, indifferent, and seems emotionally cold. Like Wölfli, he may be a loner. His range of interpersonal responses is much restricted. He has a reduced sense of purpose, and has difficulty setting himself goals and going about achieving them. In short, the schizophrenic's ability to perform acts of volition seems impaired.

The third category of symptoms involves subtle changes in patterns of perception or cognition. The effects of these changes are sometimes collectively known as 'divergent thinking' or 'loosening of associations'. Schizophrenic thought has a surreal feel about it, as connections are made which are quite bizarre from the normal point of view. When asked to sort pictures of objects into classes, schizophrenics often depart from those categories most intuitively accessible to the rest of us, such as 'tools' or 'fruit'. One subject put together a bath plug, a padlock, and a circle of red paper 'because all three stop flows or processes'. Another put together a trumpet, an umbrella, and a whistle because all of them are, in some way, 'noise-producing objects'. Schizophrenic responses can be either abnormally concrete, as when a spanner and a screwdriver are put together because they are silver rather than because they are tools, or abnormally abstract, as when a coat and a dress are put together because they both 'maintain human modesty', or air and water are put together because they are 'states of molecular density'. Either way, the perspective is unconventional.

This unconventionality can also extend into the way things are perceived. When schizophrenics are given a Rorschach test, in which the subject is asked to say what they see in a series of amorphous blots of ink, their responses are variable and surprising. For example, one patient, when asked to say what he saw, replied, 'Thick disconnected dots on the peripheral flanges of the image, what could either happen from a dripping or a smearing effect'. By contrast, another replied, 'it is an imaginary figment . . . You can consider it anthropomorphic. It's a kinship, you can identify, it's a system, belief'; and another, 'it's a symbolic volcano. It's like a volcano of thought. Thought comes out from the spirit and goes through the mind, the body, and just comes out emotionally'. Now a volcano is the kind of object commonly seen by normal subjects in a

Rorschach blot; large, concrete, significant in human life. What is strange in the schizophrenic answer is the immediate retreat into abstraction ('a volcano *of thought*'). The schizophrenic perspective is sometimes at the level of everyday objects, but is often either much more literal (the 'disconnected dots' of the blot itself) or puzzlingly abstract ('a belief system'). The schizophrenic subject will often vacillate between different scales, or even experience conflicting interpretations at the same time, as when one patient said that an inkblot reminded him of a butterfly and also of the world, hence of 'a butterfly holding the world together'. This eccentricity of interpretation of the perceptual world can be arresting, even witty, as when the schizophrenic subject passed the sign at a pedestrian crossing which said:

PED XING

and announced, 'Now we are entering the Chinese village'.

Louis Sass has dubbed the change of cognitive perspective in schizophrenia 'the loss of the middle distance'. That is to say, for the schizophrenic, the objects in the world do not always present themselves at the conventional, person-sized level of analysis. Schizophrenics describe themselves as being strangely aware that other people are composed of atoms, or of cells, or alternatively overwhelmed by the vastness of the universe while talking to someone. The oddity that makes schizophrenics conscious of literal, concrete patterns or of highly abstract categories in the Rorschach test can be thought of in the same way; the conventional 'middle distance' of butterflies and volcanoes is no more likely to come into schizophrenic consciousness than several equally possible, but usually irrelevant, alternative levels of analysis.

This loosening of association and perspective is often very damaging for everyday behaviour. However, it should not be thought of solely as a deficit. When a task requires a relaxing of normal expectations, the schizophrenic's inherent loosening is not disadvantageous. Examples of this come from experiments in which subjects have to identify stimuli that are presented in an unclear way (either a blurred picture, or a word in a sentence obscured by noise). When the picture or word is a highly predictable or conventional one (like the word *car* in the sentence 'Along the street sped a car'), schizophrenics show a performance deficit relative to control subjects. However, when the word is unexpected ('The photographer made a pretty *box*'), or the picture unconventional (a man drawing a picture with his foot), the performance of the schizophrenics is superior. Similarly, schizophrenics and individuals with highly schizotypal personalities can score highly on tasks where they have to come up with novel uses for an object, or original 'lateral thinking' connections between things. In this, their scores resemble those of the most creative normal subjects.

It is important to stress that the three groups of symptoms in schizophrenia—the positive symptoms, the negative symptoms, and the divergent thinking—have quite independent trajectories. Positive symptoms only occur during the florid phase of the psychotic break itself, and thus they may not manifest themselves until late in a person's life, and then only sporadically. Their onset may be quite sudden. It is the positive symptoms that are suppressed by neuroleptic drugs, and they may disappear entirely for ever, or at least until the next relapse. By contrast, the negative symptoms can set in years before there is a manifest psychotic break, and if they do so gradually, they are often taken as a feature of the individual's personality, not as frank disease. There is little evidence that negative symptoms are ameliorated by antipsychotic drugs, which is unfortunate, because when they are severe, they have just as much impact on the patient's life as the more obvious ravings of hallucination and delusion.

These considerations apply even more strongly to the propensity for divergent thinking. Abnormality of perspective is particularly striking in full-blown psychosis; a warning sign of an incipient breakdown is a sudden recognition that the world is vastly strange. However, it seems that divergent thinking is there in incipient form well before the breakdown. Indeed, it is a stable feature of the schizotypal make-up, since it is found in the biological relatives of schizophrenics who have not had a psychosis themselves, and also in those who score highly on schizotypy scales, but have never been ill. Both negative symptoms and divergent thinking make drawing the line between symptoms of the disease, and the traits of personality that precede the disease and predispose to it, extremely hard.

◆

What is the common thread in these diverse symptoms? One influential tradition in the psychology of schizophrenia implicates the processes of selective attention. In normal life, we are bombarded with a profusion of signals, both from the world outside us, and from the ongoing processes of our own brains. If I am trying to teach my niece to tie her shoelaces, both she and I have to home in on the relevant attributes of the situation; not the specific length of the laces, not the specific colour of the shoes, not the birdsong in the next garden, the realization that leather is made from cows, the awareness of our own breathing, our fear of heights, or the fact that we are surrounded by billions of gallons of air. Not much learning could go on if these facts all seemed equally salient. For this reason, the human mind is equipped with an attention system that allows its resources to be directed preferentially to one task or theme rather than another. Arguably a by-product of this system is consciousness itself. In our brains, the number of cognitions going on simultaneously is vast. However, our stream of consciousness is, under normal circumstances, remarkably unitary. We can tell ourselves, or others, a story of what we are thinking and doing, and though we may flit from thing to thing, there is a kind of coherence to it that contrasts strongly with the seething parallel streams of the brain itself. Most of these streams are banished from conscious thought. You are probably unaware of the weight of this book in your hands, the small shifts in posture you are making to keep yourself comfortable, that slight background noise, or your increasing anxiety about next Wednesday. Selective attention allows you to achieve this.

It is commonly observed that schizophrenics are impaired on any task that requires sustained attention to one thing. This is particularly true when there are distracting stimuli to be ignored. Schizophrenic patients are upset by background music or noise, and cannot keep these from intruding on their thoughts. Now the inability to exclude selectively irrelevant things from consciousness accounts for many features of schizophrenic cognition. In the Rorschach test, we can all see every picture at multiple levels—the literal level, as a blot, a highly abstract level, as an idea, or some intermediate level, where it resembles an everyday object. Most of us automatically exclude from consciousness all but the level that is most useful in everyday terms—that of the object—and respond that the blot looks like a bird or a whale. Similarly, most of us interact with our friends and partners without being unduly troubled by the fact that they are made of atoms, or geometrical surfaces, or that they represent the triumph of good over evil. These beliefs may be there, accessible when the context requires them, but they rarely come unbidden to mind.

For the schizophrenic, this is not always so. There is some kind of deficit in allocating consciousness selectively, and too many processes and associations—internal or external—flood to mind. This explains the positive aspects of schizotypal cognition as well as the negative ones, for it allows unconventional linkages and usages to spring to mind where the normal subject is stuck in a rut of conventional, linear thinking. So far so good, for divergent thinking, but how can the attentional theory account for positive and negative symptoms?

The British psychologist Christopher Frith has argued that positive symptoms arise because the schizophrenic misattributes things that surface in his consciousness. As we have seen, there is a failure to exclude from the mind things that are irrelevant to the situation in hand (day-dreams, hopes, fears, speculations, fragments of memory). In a psychotic break, these appear in consciousness, and once there, must be given some interpretation, for in consciousness, we are always trying to write a coherent story of ourselves. However, the schizophrenic's attention is so refracted that he cannot really monitor where they are coming from. He therefore attributes things that are generated internally to external reality, and sometimes vice versa. He experiences odd shards of internal visual imagery as being in the world; this is hallucination. He experiences fragments of cognitive processing, which would normally be ignored, as a voice in his head. Similarly, he even misrecognizes his own thoughts and intentions, and attributes them to some other force, which accounts for the very common schizophrenic experience that his thoughts are being controlled by an outside agency. In the end, the schizophrenic is still a rational human, and so he constructs large and fabulous belief systems that explain the anomalous sequences of his experiences. We saw a clear example of this with Mr Matthews in the Introduction; he had created a huge, self-consistent, and complex theory—pneumatic chemistry, republicanizing gangs, and so on—which neatly explained all the bizarre things that came to him and which he misattributed to reality.

This theory is satisfying because it predicts both the enormous variability of positive symptoms, and some common threads. It predicts variability because everyone has a different set of memories, perceptions, and relationships to go into the flux of cognition. Thus we would predict that hallucinations would be no more similar from person to person than the human imagination in general, and this is quite true. No two patients have positive symptoms with exactly the same content.

On the other hand, there are some recurrences of form in positive symptoms which are consistent with the misattribution idea. Influencing machines, thought control by the CIA, and a running voice commenting on one's own behaviour, are all examples of misattributing one's own intentions to something else, and believing that some unrelated event has a special significance for oneself is an example of misattributing an unconnected fact to one's own state. The common schizophrenic loss of the boundary between the self and the world reflects this confusion.

How, though, can the ideas of failure of selective attention and attribution be applied to negative symptoms? Well, if Frith is right that the schizophrenic has trouble identifying and attending to his own acts of volition, this would account for the schizophrenic deficit in initiating and completing goal-directed activities. Flatness of emotion, too, could come from a failure to identify correctly and exclusively one's own feelings towards another person. This idea is borne out by the fact that the schizophrenic's experience of emotion is often not that of the lack of emotion, but rather of inappropriate emotion, or conflicting feelings all at the same time. Social withdrawal may well be the person's way of adapting to the painful and confusing experiences that social interaction must offer when one cannot get one's attention and feelings quite straight.

The selective attention hypothesis, then, provides an explanation of the mechanisms of schizophrenia at the psychological level. But what is the basis in the brain for these psychological processes?

◆

Research into the brain mechanisms underlying schizophrenia yielded no significant fruits until the late 1950s, after the antipsychotic drugs reserpine and chlorpromazine had become available. Both these drugs were subsequently shown to suppress the action of the monoamine neurotransmitter dopamine. Soon afterwards, it was shown that drugs like cocaine and amfetamines, which can produce a temporary psychosis virtually indistinguishable from the positive symptoms of schizophrenia, enhance dopamine function. These observations lead to the so-called dopamine hypothesis of schizophrenia.

In its simplest form, the dopamine hypothesis states that schizophrenic symptoms are the result of an excess of brain dopamine activity. Initial evidence on whether this was indeed the cause of schizophrenia was disappointing. The postmortem schizophrenic brain was not consistently found to contain elevated levels of dopamine or of dopamine activity's chemical end-product, homovanillic acid. Only much more recently has it been shown, using PET scanning, that the rate of certain chemical reactions related to dopamine production is abnormally high in parts of the schizophrenic brain. However, the relationship between the rate of these reactions and the absolute abundance of dopamine at the synapse is not yet clear.

The dopamine hypothesis switched to a slightly different tack when it was discovered that the density of a particular class of receptor molecule (D_2) is consistently high in schizophrenics examined at post mortem. This led to a slight reformulation of the dopamine hypothesis. Rather than the transmitter being too plentiful, perhaps the receptors are super-abundant. In fact, this superabundance could actually result from the long-term *under*-supply of dopamine to brain circuits. The brain would then up-regulate the production of receptor to compensate. One effect of this up-regulation would be that when there was any temporary surge in the level of dopamine, the sufferer would be super-sensitive to its effects.

This picture makes a lot of sense. The negative symptoms of schizophrenia would be due to long-term under-activity of dopamine circuits related to motivation and reward, whilst the intermittent positive symptoms would be excessive responses to periodic surges of dopamine activity. Such surges would do to the dopamine-sensitive schizophrenic what large doses of amfetamines would do to anyone else. This view also explains why antipsychotic drugs which damp down dopamine activity suppress positive symptoms but leave negative symptoms untouched.

Unfortunately for the D_2 receptor hypothesis, it transpired that one of the effects of conventional antipsychotic drugs is to increase the density of dopamine receptors in the brain. This happens because the drugs blockade the receptors, in response to which the brain up-regulates production. It could be, then, that the increased receptor density found at post-mortems simply reflects the effects of long-term medication. The crucial question thus became whether receptor density is increased in the drug-free schizophrenic brain. This question became tractable in the 1980s with the development of PET scanning techniques that allowed receptor densities to be measured in the working brain, rather than just at post-mortem. A number of studies using these techniques on 'drug-virgin' schizophrenics have now been published. The results are not unanimous, but the general finding has been that there is a modest but significant increase in D_2 density compared with normal controls.

The D_2 hypothesis, then, is working along the right lines. The immediate neurochemical problem in schizophrenia is some kind of imbalance between transmitter and receptor in parts of the dopaminergic system. The situation is undoubtedly more complex than a simple excess of D_2 molecules, though. A recent study by Anissa Abi-Dargham and colleagues from Columbia University suggests that the abnormality is in the proportion of D_2 receptors actually occupied by dopamine at any one time. This proportion appears to be higher in schizophrenic than in normal brains.

As well as D_2, other receptor sub-types are probably implicated in the imbalance, and there is specificity to particular brain areas. Some post-mortem studies show an excess of dopamine in the amygdala of just the left hemisphere of the schizophrenic brain. This may well be a significant finding, since functions are not shared equally between the two hemispheres, and the left hemisphere is considered dominant in most people for language and consciousness. Indeed, there is some evidence that normal left hemisphere dominance is absent or attenuated in schizophrenia.

Another avenue of research concerns the interaction between dopamine and other neurotransmitters, particularly glutamate and serotonin. Glutamate is implicated in the excitation of large parts of the thinking brain, and is partly regulated by dopamine. Serotonin, as well as its well-known relationship with antidepressants, is strongly affected by antipsychotic drugs. Some of these drugs are thought to have their main effects through serotonin rather than dopamine, and what is more, hallucinogens such as LSD resemble serotonin more than dopamine. A PET scanning study has recently shown that the concentration of one class of serotonin receptor is reduced in the brains of a group of 'drug-virgin' schizophrenics.

The details of dopamine dysregulation in schizophrenia thus await further clarification, but it is clear that dopamine is crucially implicated in the disorder, and equally clear that it does not operate simply, or in isolation.

As the dopamine hypothesis has become more complex, another strand of research on the mechanisms of schizophrenia has come further to the fore. This strand concerns not the chemistry but the actual structure of the brain. It was spurred by the finding, first published in 1976 and subsequently confirmed, that schizophrenics have an enlargement of the hollow spaces deep in the brain known as the cerebral ventricles. Such an enlargement indicates atrophy of the surrounding tissue in the mid-brain.

Following this result, a large number of studies of the schizophrenic brain were carried out, using both scanning and post-mortem techniques. Unfortunately, the results were not always consistent, although the finding of increased ventricular size was fairly reliable. Some, but not all, reports have found schizophrenic brains to be smaller and lighter overall, whereas several have found more specific areas to be reduced. The regions involved include the thalamus, and the temporal and frontal lobes of the cerebral cortex. The amygdala and hippocampus, structures lying underneath the temporal lobe and responsible for emotion and episodic memory, respectively, have often been found to be smaller, particularly on the left side, though again this result is not completely consistent. The cortex above the hippocampus has also been found to be abnormally thin in schizophrenics. There are also reports of odd patterns of cell arrangement, size, and density in several brain areas, including the hippocampus and the surrounding cortex, parts of the frontal lobes, and parts of the thalamus. Once again, though, many of the reports await consistent replication.

These results might appear somewhat unsatisfying. Apart from the consistent ventricular enlargement, the findings involve a large number of different brain abnormalities, and a poor degree of consistency between subjects and between studies. Indeed the problem seems to be not so much a paucity of differences between schizophrenics and normals, but an embarrassment of riches, with every study finding something slightly different. In many ways, this mirrors the central problem of schizophrenia itself, which is the heterogeneity of the phenomenon. Surely the evidence that schizophrenia is a brain abnormality would be more convincing if there were a single, discrete difference that turned up in every case?

However, further consideration of the nature of the disorder reveals that no such locus is likely to be found. I have argued that what is impaired in schizophrenia is not any single function—like emotion, perception, memory, or language—but rather the effective integration of all the functions into a single, coherent stream of consciousness. Brain scientists believe that different brain areas are specialized for different functions, and that the processing of these functions occurs basically in parallel. The achievement of a unified supervisory beam of conscious attention, by contrast, does not happen in any one place, but rather concerns the way that different, widely distributed brain circuits interact. Therefore, any brain differences involved are not going to be highly localized, but rather will be abnormalities in the patterns of connectivity between various areas. The anatomical differences observed fit well with this pattern. The hippocampus and the areas around it, for example, integrate information from all over the brain, and the thalamus is a kind of relay station for all of the cerebral cortex. The inconsistency of the results is also interpretable from this perspective. The developing brain is highly plastic, and self-organizes to a very considerable degree; a small abnormality in its initial developmental programme would have results that varied in detail from individual to individual, but always involved oddities of neuronal arrangement and connectivity; and this is indeed what the anatomical studies find.

This view—that the pathology in schizophrenia is in the connecting up of parts, rather than in any one part—finds support from studies using functional brain scanning techniques such as PET. PET measures the metabolic activity of different parts of the brain in awake, behaving subjects in real time. Early PET studies of schizophrenics suggested an underactivation of the frontal lobes. This pattern is known as hypofrontality. Hereafter, the hex of schizophrenia research set in once again, with some subsequent studies confirming the pattern, and some not. However, these studies investigated mainly frontal lobe metabolism with the subject resting. More interesting are the studies where the subject had to undertake a cognitive task.

When a normal person takes on a task where executive, rational decisions are required, there is a diversion of cerebral blood flow towards the frontal lobes, and away from other areas of the cortex. This is because the frontal lobes, which are much enlarged in the human brain, are specialized for such functions. When schizophrenics are presented with such a task, the increase in frontal activity is not as much as it is in the normal subject, especially on the left. At the same time, there is not the reduction in activity in another brain area, the superior temporal cortex, that is normally seen. Patterns of activation in the cerebellum and thalamus are also abnormal relative to the task in hand. In general, then, the results of PET scanning point the same way as the results of anatomical studies; there is no one area where activation is extreme, but rather a failure of widely distributed areas to coordinate their firing in the usual way. In general, it seems that there is a dysregulation of the relative activity of different parts of the brain.

◆

We seem to have two radically different stories about the mechanisms of schizophrenia. In one, the dopamine hypothesis, the dysfunction is a chemical one. An imbalance of a neurotransmitter (or several, given the interactions of dopamine with glutamate and serotonin) leads to an odd pattern of activity in the brain circuits served by those transmitters. But it is assumed that the circuits are basically there, if they only had the right chemical software to operate them. The other story is an anatomical one. It stresses differences in the way the different areas of the schizophrenic brain are actually wired up. In the anatomical story, schizophrenia is a hardware rather than a software problem.

In fact, the distance between these two stories is not as great as it might seem, for in the brain the hardware and the software constantly interact. Different brain circuits are served by different sets of neurotransmitters, so if those circuits are damaged, there will be an imbalance of the transmitter and its products. On the other hand, a deficiency in a transmitter in the developing brain will lead to atrophy of the circuits that would normally use it. The enlargement of the cerebral ventricles in schizophrenia is related to the atrophy of cells in the surrounding tissue, tissue which, in the normal brain, is the base of several key dopamine pathways to the cortex. The chemistry and the anatomy are thus bound up together, and it is hard to say which one comes first.

That said, the anatomical differences are clearly prior in schizophrenia in some important sense. The dopamine overactivity hypothesis is only really an account of the positive symptoms of schizophrenia, not the whole syndrome. It is the positive symptoms that are suppressed by dopamine-blocking drugs, and which are mimicked by dopaminergic drugs such as amfetamines. The other symptoms of schizophrenia, the underlying oddities of attention and cognition, are much more difficult to treat, and may come on years before any overt positive symptoms appear. It is probably better to say, then, that the ultimate mechanisms lying behind schizophrenia are oddities in structure of the brain, related to an abnormal pattern of coordination and connectivity between different areas. These oddities cause, amongst other things, a vulnerability to periodic dysregulation of neurotransmitter systems. Such a dysregulation, when it flares up, produces the positive symptoms of psychosis. Antipsychotic drugs tackle this, but leave the underlying anatomical basis largely untouched. The causes of these brain abnormalities are, for the most part, presumed to be oddities in the genetic programme which sets off the development of the brain.

If this view is correct, then there is an asymmetry between affective psychosis and schizophrenia. In affective psychosis, the most striking symptoms—changes in emotional tone—are almost completely reversible if the right neurochemical key is found. Although aspects of the manic-depressive temperament persist through treatment (and there is some evidence for permanent brain changes, such as ventricular enlargement), most of the danger can be taken out of the disorder using lithium.

Schizophrenia, by contrast, seems to be more structural. This point is borne out by the different courses of the two conditions. For schizophrenics, the relationship of the disease to life stress is less obvious, the long-term prospects rather bleaker, and the effectiveness of drugs and psychotherapy in controlling the disorder somewhat less than is the case for affective illness. None the less, we should not be too bleak about the outlook. Schizotypal individuals only rarely become schizophrenic, and even when the break does occur, there are plenty of people who recover completely, as we shall see in a later chapter.

♦

When discussing affective psychosis, we found a normal precursor for the disorder in the mood system. The mood system, I argued, was universal and entirely adaptive, and individuals vulnerable to affective disorder were just those at the ends of the continuum in the reactivity of this system. They were so far along the scale that their moodiness had become pathological. Can we make parallel generalizations about schizophrenia?

I have already endorsed the view that there is a continuum of schizotypal personality, which represents a susceptibility to divergent thought and odd experience. Being highly schizotypal may or may not lead to a break into the positive symptoms of schizophrenic psychosis. But what is the normal function of schizotypal cognition?

I have argued that in order for the mind to function effectively, there must be some purposeful integration and prioritization of the many cognitive activities that are going on in parallel. This integration is referred to as selective attention, and one of its side-effects is consciousness. The attentional beam must, by definition, filter out bizarre and divergent associations, and irrelevant stimuli. However, it should not filter them out too effectively. That is to say, the mind that cannot jump from track to track, or make a leap of creative insight, or find a non-obvious connection between two different domains, is at a great disadvantage compared to one that can. A little divergent thinking is surely a great advantage in many activities. Just as with moods, the problem is one of balance. Too much divergent thinking leads to cognitive disorganization, eccentricity, and eventually to delusion and hallucination, but a little is highly beneficial. In fact, even a large dose can be put to some striking and original uses, as Adolf Wölfli's output so clearly demonstrates.

This view of the benign functions of schizotypy has a long history. It was espoused by the influential Victorian psychiatrist Henry Maudsley in 1871:

> I have long had a suspicion . . . that mankind is indebted for much of its individuality and for certain forms of genius to individuals [with] some predisposition to insanity. They have often taken up the by-paths of thought, which have been overlooked by more stable intellects . . . There is sufficient truth in the saying 'men think in packs, as jackals hunt' to make welcome in any age . . . the man who can break through the usual routine of thought and action.

The evidence for a link between schizotypal traits and creative thinking is quite strong. As I have already mentioned, on tasks that require novel connections and associations, schizotypal individuals and very creative people score equally highly. Now the positive symptoms of schizophrenia are often so bizarre as to impair sustained work, while the negative symptoms destroy the motivation to do so, but the other facet of the illness, the capacity for divergent thinking, is the common core of both intense creativity and madness. Even in the midst of the schizophrenic breakdown, there are flashes of this core, which subjects often describe in terms of a strangely heightened awareness of the world and of imaginative connections within it. Adolf Wölfli used these in his art, as we have seen, and another schizophrenic artist, M. E. McGrath, describes her moments of creative insight in the depths of the storm in the epigraph to this chapter.

In fact, creative individuals have often sought to cultivate something very close to the schizotypal experience as a way into their work. The French modernist poet Arthur Rimbaud proclaimed that poetry must proceed by a 'systematic derangement of all the senses'. Giorgio di Chirico, another artist whose work pushed twentieth-century painting firmly in the direction of surrealism, was severely schizotypal, if not schizophrenic. Like Wölfli, he used the oddness of perception he experienced as a springboard for artistic innovation:

> I saw then that every angle of the palace, every column, every window had a soul that was an enigma . . . Then I had the strange impression that I was looking at all these things for the first time, and the composition of my picture came to my mind's eye. Above all, a great sensitivity is needed. One must picture everything in the world as an enigma, not only the great questions one has always asked oneself . . . But rather to understand the enigma of things generally considered insignificant . . . To live in the world as if in an immense museum of strangeness.

The schizophrenic world is indeed an immense museum of strangeness, of the profound, imaginative, disturbing strangeness of which the human mind is so uniquely capable.

CHAPTER 6

Such tricks hath strong imagination

> The only difference between me and a madman is that I am not mad.
>
> Salvador Dali

If I have been doing my job well, the preceding chapters will have convinced you of four propositions. Firstly, psychosis is a reasonably common part of human experience, affecting something like 1 in 30 people in a typical population. Secondly, the major psychoses—schizophrenia and affective psychosis—consist of severe disruptions of cognition and emotion, which have a basis in the brain. Thirdly, the consequences of these disruptions are personally disastrous. They come on in the prime of life, and frequently lead to suicide, the loss of intimate relationships, and the loss of one's capacity to provide for oneself. One can imagine no more severe impairment of social success than a major psychosis, short of actually dying. Fourthly, the liability to these disruptions is dependent on personality and, like other personality factors, it is substantially inherited in the genes.

These four statements, taken together, add up to an intriguing puzzle. If there are genetic variants that predispose to psychosis, and if psychosis has terrible effects on such life outcomes as reproductive success, then why have those genes persisted in the human species? Why hasn't the firm hand of natural selection ejected them from the gene pool? The genetic variants that predispose to psychosis would seem, from an evolutionary point of view, to be a ticking bomb, ready to blow themselves and the genes around them to biological oblivion by destroying the functioning of the individual who carries them.

Now if there were evidence that psychosis was very rare, had arisen only very recently, or was restricted to a few small populations somewhere rather remote, then I could find an easy solution to the puzzle. That answer would run as follows: the genetic variants underlying psychosis are nasty mutations that have cropped up recently in some place or other. Since these mutations do not always express their effects—that is to say, the bomb does not always go off—they sometimes get passed on to the next generation, and natural selection, which works by gentle, gradual changes in gene frequencies, hasn't had time to weed them out yet. But it will.

The problem with this story is that the distribution of psychosis does not bear it out. Psychosis was known to the ancients of all cultures, and has persisted right through to the present day. It is observed, as we shall see later, in every human population that has ever been studied, probably at fairly comparable rates. Furthermore, the genetic variants related to it are anything but rare. At least one copy of the short version of the serotonin transporter regulatory gene, which appears to move its bearer a step towards liability to mental disorder, is found in 68 per cent of the population. This is hardly the profile of a recent, rare, nasty mutation.

I have argued that no single gene is responsible for the risk of psychosis. Rather, there are dozens of genes that jointly determine our place in a normal distribution of psychoticism, and the extreme personalities are the ones at risk. Perhaps this is the answer to our puzzle; since no gene individually accounts for more than a tiny fraction of the risk of psychosis, there is no single gene that can be selected out as the culprit.

Unfortunately, this answer does not work either. Natural selection works on traits that are genetically complex just as inexorably as it does on traits controlled by a single gene. As long as a genetic variant is statistically associated with a negative trait, even if it can only determine that trait in conjunction with many other genes, then selection will eventually weed it out. Geneticists have a good understanding of how this weeding out affects the gene pool. Let us once again take height as an example of a polygenic trait. Imagine a short, necky kind of creature, which at some point in its evolutionary history starts to procure all its food from the low branches of a particular tree species. With astonishing convenience for us, all the trees of this species are exactly the same height. Because there are many genes affecting the height of the necky creature, there will, at the beginning of our thought experiment, be a normal distribution of heights in the population. Initially, the tallest individuals, closest to the larder, will do very well and outreproduce the shorter ones. There will thus be more offspring of tall parents in the next generation. This is *directional selection*, and it will implacably pull the mean height of the animal up until it is precisely the same as the height at which the food is to be found.

Once this mean height has been reached, there is no advantage to being any taller, since bending down to reach food is about as enervating as craning upwards. The mean height in the population will stay the same. However, those at *either* extreme of the height distribution will now be penalised, and their representation in the gene pool diminished. This is *stabilizing selection*, and its effect on the population is to keep the mean of the trait the same, while pulling the bell curve ever narrower around that mean. In fact, if there are no other selective forces, all trees stay the same height, and the population is large, then in the end, all individuals will be the same as regards the genes controlling height. The bell curve will be very narrow, with any residual variation in height due to the environment and not to genes.

Let us apply this logic to the personality axis of psychoticism. Recall my argument that this can be decomposed into two sub-axes of thymotypy and schizotypy. I argued that a

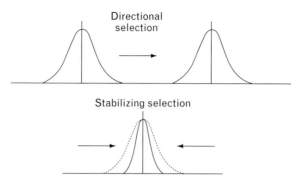

Figure 8 The effects of directional and stabilizing selection on a continuous, polygenic trait such as height or personality.

little thymotypy is a good thing. Moods are, after all, useful, in that they guide us to appropriate life goals. Similarly, a little schizotypy is a good thing, as the capacity for creative or divergent thought within a basically focused consciousness is invaluable. For both of these traits, then, there is an optimum level, well above zero, but apparently below that found in psychotics. Directional selection can be relied on to move the average for the human population to those optimum levels, and we can reasonably assume it has done so. However, very significant genetic variation around this optimum is still found, which is one of the things that makes human life such a rich pageant, and also what accounts for the persistence of psychosis. We know that at one extreme of the distribution, there are terrible negative effects. The question therefore becomes: why has stabilizing selection not narrowed the bell curve of psychoticism, either down to zero or down to the point where no psychosis ever occurs? Why are there still so many people with genotypes high in psychoticism in the human population? Why are we, as a species, more crazy than is good for us?

We are now head on against the central question of this book. It was a question already posed, in general terms, in the Introduction, but the things we have learned along the way have allowed us to formulate it much more precisely. In the Introduction, I discussed Shakespeare's answer. He had Theseus declare that:

> The lunatic, the lover and the poet
> Are of imagination all compact.

That is to say, the traits that underlie psychosis also underlie the vision of the poet. The essence of these traits is identified specifically as strong imagination. The suggestion is that strong imagination is a double-edged sword; as well as its fearful effects in psychosis, it has beneficial ones in creativity. The selective advantage of these beneficial effects would counteract the selective disadvantage of psychosis. Thus, genotypes high in psychoticism would remain in circulation in the human population. This would explain why we still have the capacity for madness. In this chapter, we shall see how far-sighted Shakespeare's vision turned out to be.

◆

There is a long cultural tradition, revived from time to time, of relating madness to creative genius. The position has a strong intuitive appeal, and it is easy to find examples of individuals who seem to fit the profile. However, intuitions are there to be mistrusted, and swapping anecdotes proves nothing. After all, psychological studies of the creative mind have shown again and again that, however much we might want to romanticize it, it is typified by qualities that are disappointingly opposite of psychotic; self-discipline, tenacity, organization, calmness, and strong self-image. This is the conundrum of creative achievement; is it about fiery leaps of inspiration, or a banal, Calvinistic rationing of perspiration? Or both? What we need in order to approach this question in a scientific way are controlled statistical studies of the association between psychotic and creative traits.

Such studies have been carried out, and the results are overwhelmingly consistent. Whatever other qualities there might be, there is very often a touch of fire in the mind. Kay Redfield Jamison studied biographical and autobiographical material on all major British and Irish poets born between 1705 and 1805. She paid particular attention to suggestions of symptoms of mania or depression, and the patterns of mood, energy, and work. This was not a properly controlled study, since Professor Jamison knew the hypothesis in advance and went looking for evidence. The diagnoses she made were not independently verified. Indeed, they could not be, since the individuals in question had all died before meaningful psychiatric diagnosis existed. None the less, Jamison knew what she was looking for, and provided ample evidence for her conclusions.

She found a strikingly high incidence of mental problems. Six of the 36 poets were committed to lunatic asylums. Another two committed suicide. More than half showed strong evidence of mood disorder, either bipolar or unipolar, many including overtly psychotic symptoms. Many of the poets had family histories of mental disorder or suicide. Taken together, the results suggest that to be a poet in Britain in the eighteenth century was to run a risk of bipolar disorder 10–30 times the national average, suicide 5 times the national average, and incarceration in the madhouse at least 20 times the national average. Hardly an appealing prospect, when you think about it.

In a subsequent study, Professor Jamison extended her research to living creators. She obtained the cooperation of 47 poets, writers, and visual artists who had won major prizes or awards in their fields. Because she was working with a contemporary population this time, rather than devising her own diagnosis, she could record the diagnoses and treatments the subjects had received from their physicians prior to the study.

Thirty-eight per cent of the subjects had received treatment for an affective disorder. Twenty-nine per cent had taken antidepressants or lithium, or been hospitalized. The poets and playwrights had the highest rates: 55 per cent of the poets and 63 per cent of the playwrights had a diagnosis of mood disorder. This compared to around 20 per cent for novelists, biographers, and artists. Even this is huge compared to the comparable rate for the general population, which is perhaps 6 per cent if we consider those that meet the diagnostic criteria, and significantly less if we consider those who actually seek medical help. Which means, once again, that poets have at least 10 times the going rate for affective disorder.

These results have been confirmed substantially by investigations carried out by Nancy Andreasen into members of the prestigious Iowa Writers' Workshop in the USA. Dealing once again with living professional writers, Professor Andreasen was able to devise and conduct standardized diagnostic interviews. Her study was also valuable for including a control group of people matched for age, sex, and educational attainment, but not engaged in the creative arts. She found a staggering 80 per cent of the writers qualified for a diagnosis of affective disorder, compared to 30 per cent for the controls. This control figure is surprisingly high. It may reflect something about the social group from which the controls were drawn, or the generosity of the diagnostic criteria used. None the less, the difference between the writers and the controls is highly significant.

All these studies have been restricted to creative writers and a few visual artists, and they have consisted of relatively small samples. They do not prove that there is something about creativity, as opposed to, say, fame, high intelligence, or a stressful profession, which is particularly associated with mental disorder. What is needed is a large sample of eminent people across many professions, allowing a comparison of rates of psychosis.

The gargantuan endeavours of Professor Arnold Ludwig, reported in his book *The Price of Greatness*, provide just such a sample. He collected biographical material on 1004 eminent men and women. The criterion for eminence was that the subjects had a biography published between 1960 and 1990, and reviewed in the *New York Times Book Review*. The dates and the choice of journal are arbitrary, but they provide a sampling framework for the study. For each of his eminent people, Ludwig assiduously recorded information on lifetime achievement in their field, family history, health (including mental illness), sexuality, and behaviours such as suicide, alcoholism, and drug addiction. This mammoth task took him through over 2200 different biographies.

This study suffers from certain limitations. The diagnoses are provided retrospectively by Ludwig himself, not by an independent physician, although of course the testimony of physicians attending the eminent people may be recorded in the biographies. Furthermore, there is no independent control group of non-eminent people, so we cannot look at the effects of being eminent per se except by comparing rates for the population at large, and this is unsatisfactory, since the eminent people were certainly not matched to the population at large for intelligence, educational attainment, or stress. However, the study is invaluable for asking other types of questions, because of the vast sample, and because it allows the direct comparison of people who were eminent in creative professions and those eminent for other reasons.

Ludwig found a lifetime prevalence of any psychiatric disorder of 59 per cent amongst his eminent people. The relevant comparison to the population at large is difficult to make, since his criteria were generous, and general population studies using such a broad brush have sometimes reported rates almost this high. However, the comparisons between the professions are very instructive. Prevalences of disorder in business (49 per cent), exploration (27 per cent), public office (35 per cent), natural science (28 per cent), and the military (30 per cent), pale into insignificance when compared with those observed in creative pursuits: 87 per cent for poets, 77 per cent for fiction writers, 74 per cent in the theatre, 60–68 per cent in music, and 73 per cent in the visual arts. The increased risk in creative pursuits is strikingly clear, with an average risk of 73 per cent for the creative arts, compared to 42 per cent for all other callings.

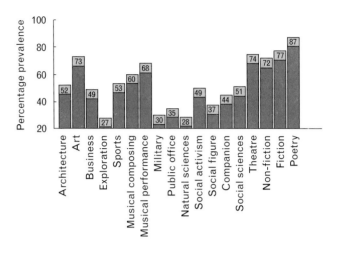

Figure 9 Lifetime rate of any mental disorder in 1004 eminent people, classified by profession. (From Ludwig A. M. (1995). *The Price of Greatness*. © 1995 by Arnold M. Ludwig. Reprinted with the permission of Guilford Publications, Inc.)

Ludwig also broke down the mental disorders by category, and this analysis revealed that the elevated rate in creative professions was not restricted to any single psychiatric symptom. Creative professionals had higher rates of depression (50 per cent versus 24 per cent), mania (11 per cent versus 3 per cent), severe anxiety (11 per cent versus 5 per cent), and suicide (15 per cent versus 5 per cent, compared to a general rate in the US of about 1 per cent) than the eminent people in other walks of life. Ludwig's study also allowed the consideration of schizophrenia as well as affective disorders. The studies by Jamison and Andreasen previously discussed show no evidence for a schizophrenia–creativity connection. This does not mean that there isn't one; Jamison went looking specifically for affective disorder, and Andreasen's sample consisted of only 30 people. The absence of any schizophrenia may just reflect the limitations of the sample. Ludwig's sample, by contrast, should be big enough to capture any association between schizophrenia and creativity.

Unfortunately, establishing an unequivocal diagnosis of schizophrenia is not easy when one is dealing with biographical material from periods when psychiatric diagnosis was less rigorous than it is today. Cases described at the time as schizophrenia may actually be better characterized as acute mania, psychotic depression, or schizoaffective disorder. To avoid the difficulty of drawing a sharp line, Ludwig scored his subjects for the presence of any schizophrenia-like psychosis, including such positive symptoms as delusions and hallucinations, with the understanding that this is a somewhat broader category than schizophrenia as currently conceived.

Schizophrenia-like psychosis showed a similar pattern to the other disorders. The overall prevalence in the sample was 5 per cent. It was higher in the creative arts (7 per cent) than other professions (3 per cent), and particularly concentrated in poets (17 per cent), prose writers (7–8 per cent), the theatre (6 per cent), composers (10 per cent), and, perhaps surprisingly, sportsmen (11 per cent). It was absent altogether from the explorers, military officers, and public servants.

Ludwig's biographical approach has also been replicated by the British psychiatrist Felix Post. Post examined biographies of 291 world-famous men who had made contributions in six different categories (scientists, composers, politicians, visual artists, thinkers, and writers). He scored them for the presence or absence of various psychiatric symptoms. Post's study is interesting because of his stricter attention to current diagnostic categories, and because he distinguished the severity of disorder each person experienced.

Although there are some different patterns, and the sample is much smaller, many of Post's results confirm Ludwig's. All the eminent people had a fairly high rate of underlying psychological abnormality, around 60 per cent. In 34 per cent of them, this had at some time or other developed into an acute psychiatric condition, which had at least temporarily interrupted their professional activity. The lowest rates of psychopathology were amongst the scientists (52 per cent), and the highest amongst the creative writers (90 per cent).

These results are very clear. There is an increased risk of psychosis and related disorders in those who become eminent in the creative arts. This confirms my suspicion that once you go looking for psychotic traits in the creative élite, you can produce an impressive list. In poetry, we have Baudelaire, Robert Brooke, Byron, Coleridge, T. S. Eliot, Keats, Sylvia Plath, Pope, Robert Lowell, Ezra Pound, Shelley, Dylan Thomas, Tennyson, and Walt Whitman, to name but a few. In prose and dramatic writing, the list would have to include J. M. Barrie, Joseph Conrad, Noel Coward, Charles Dickens, Fyodor Dostoevsky, William Faulkner, F. Scott Fitzgerald, Nikolai Gogol, Ernest Hemingway, Hermann Hesse, Victor Hugo, Henry James, Samuel Johnson, James Joyce, Franz Kafka, Immanuel Kant, Guy de Maupassant, Herman Melville, Marcel Proust, Jean-Jacques Rousseau, Robert Louis Stevenson, August Strindberg, Leo Tolstoy, Evelyn Waugh, Tennesee Williams, Mary Wollstonecraft, and Virgina Woolf. In the sphere of music, we should not neglect to mention, alongside Schumann, Beethoven, Berlioz, Bruckner, Chopin, Dowland, Elgar, Handel, Holst, Mahler, Rachmaninov, Rossini, Tchaikovsky, and Wagner. Finally, amongst visual artists, it would be amiss not to identify at the very least Borromini, Cézanne, di Chirico, Gaugin, Goya, Van Gogh, Kandinsky, Michelangelo, Modigliani, Munch, Picasso, Jackson Pollock, and Mark Rothko. In fact, once a list like this is begun, it is hard to avoid the conclusion that most of the canon of Western culture was produced by people with a touch of madness. This conclusion would seem to confirm the hypothesis that the compensatory benefit that keeps the extremes of psychotic personality in the human gene pool is enhanced creativity. Some caveats are in order, though.

The first of these is that all these studies selected their sample on the basis of achieved eminence in a creative activity, as endorsed by winning a prize, being part of a prestigious writing group, or having a biography published. Such studies do not demonstrate an association between psychotic traits and creative *capacity* so much as an association between psychotic traits and creative *recognition*. This may reflect something about what contemporary Western culture chooses to bestow value on. We live, after all, in the shadow of the Romantic movement, and our very model of what is creative excellence is infused with their ideas about strong emotion and nonconformity. Obviously, such cultural values bias us towards recognition of the traits of the psychotic, and one has to ask whether the view of the link between creativity and madness would be different if we were living in the Renaissance, or Ancient Greece, with their greater emphasis on balance, naturalism, and perfection of form. The inclusion of Handel and Michelangelo in the list would seem to argue against this, but the point should be considered.

One study addresses this problem by approaching the issue the other way around. Instead of taking creative people and looking for evidence of psychosis, Ruth Richards and her colleagues took psychotics and looked for evidence of creativity. Specifically, they took a sample of adoptees suffering from severe or mild forms of bipolar disorder, paired them with a sample of controls matched for age, sex, and socio-economic status, and scored them on creativity scales. These scales were based not on public acclaim but on evidence of significant originality in real-life activity, whether it fell in a glamorous field or not. For example, someone who introduced new products in the chemical industry, started a major company, and found a way of smuggling explosives to the Danish resistance during the Second World War scores just as highly for lifetime creativity as a recognized choreographer. This scale probably taps the core of creativity, which is important across time and cultures, more significantly than do achieved accolades. The study found that overall creativity scores were higher for those with affective disorder than the controls, but only slightly so.

These results are suggestive, but do not completely solve our puzzle. This is because of the second caveat raised by the studies I have reviewed so far. It may be true that a few remarkable individuals have been able to ride their psychoticism into enormous, historic creativity. However, their number, even if it runs to a few hundred or thousand over the centuries, is insignificant compared to the millions of people for whom psychosis is an entirely disorganizing, horrible, unredeeming canker. To pluck the flower of art from the nettle of psychosis takes unusual intelligence and discipline, and most people high in psychoticism do not possess this. Instead, psychosis stings them to an early and unhappy grave. In fact, even amongst the very few who find a way out into artistic creation, it is not obvious that reproductive success is heightened, for though these people are culturally attractive, they are often unstable, unable to maintain social bonds, and have a much heightened risk of violent early death. To argue that the creativity I have described is enough to compensate for the psychosis at any level may be to over-romanticize. Psychosis is too awful for the story to ring quite true.

Hold on, though, for there is a further twist to the tale. The genes I am trying to explain the persistence of are not genes for psychosis. They are genes for psychot*icism*. That is to say, the genes produce the underlying personality type, just one of whose features is a risk of psychosis. As we have seen from twin studies, only about half the people with the genotype develop the psychosis. The other half carry all the genes onwards without that debility. What we should be looking for, then, is not just evidence of a compensatory benefit in the lives of psychotics, but of a compensatory benefit in the well relatives of psychotics. These are the people most likely to be taking the genes forward.

Early evidence that the relatives of psychotics might be the key came from a study by Heston in 1966 on twins from schizophrenic parents, who had gone to new families for adoption. Heston was interested in the nature–nurture question, so his prime concern was whether the twins showed the colour of their genes and became schizophrenic, or of their environments, and remained well. In fact he found strong evidence for genetic factors. However, as in other studies, if one twin became schizophrenic, the other one had only about a 50 per cent risk of doing so, despite being monozygotic and therefore carrying all the same genes. Heston noticed that the healthy twins often went on to be successful and unorthodox people, with a particular concentration in creative pursuits. It was as if the genotype led down a road with a junction in it. At the junction, you could toss a coin: heads one way, to a successful life outcome; tails the other, this way madness lies.

This view was subsequently confirmed by painstaking work on the distribution of psychosis in Iceland by Jon Löve Karlsson of the Institute of Genetics in Reykjavik. Iceland is an ideal place to study the descent of genes over time. The population is small—around 200 000 people—and stable over time, as it is a very long way to anywhere else, and there are few migrants in or out. There are excellent demographic records, and only one mental hospital, which makes tracing psychotic families a relatively easy task.

Karlsson took people with a history of psychosis as his index cases, and looked at the achievements of their first-degree relatives. He found more published authors, holders of doctorates, professors, parliamentarians, and clergymen amongst the relatives than could be expected by chance. The relatives were 30 per cent more likely than the general population to be listed in the Icelandic version of *Who's Who*, and 50 per cent more likely to have published a book. More than double the expected number were involved in the arts or scholarship. Furthermore, when Karlsson looked at large family trees, he found that the branches in which psychosis was found were the same ones as those containing many eminent people. If that were not enough, other studies point the same way. Nancy Andreasen found a rate of mental disorder of 42 per cent in the relatives of the writers in the Iowa Writers Workshop, compared to 8 per cent for the relatives of controls. As we saw from Ruth Richards' study, manic-depressives score slightly higher on creativity scales than controls. But the *unaffected* relatives of manic-depressives score higher still. These are the people reaping most of the compensatory benefit, and we have to assume they do this because they profit from the more positive traits of the disposition without the life-long disorganization of falling down the slope.

All these results suggest that the genes associated with psychosis confer a creativity benefit not just on psychotics but on their well relatives. One final study clinches it by ruling out the effects of nurture in this nexus. Thomas McNeil studied rates of creative achievement in Danish subjects who had been adopted away from their parents within weeks of birth. He divided them into high, medium, and low creative achievement categories. He found not only that rates of mental illness were highest in the high creativity group, but that the rates of mental illness in their biological parents were elevated too. There was, by contrast, no evidence of increased mental illness in their adoptive parents.

♦

So what are the traits in the psychotic personality that enhance creativity? For the affective disorders, it is clear that the ability to induce a sustained high mood is the key. We have already seen how Schumann used his high mood to compose an astonishing quantity of music, and he is not an isolated example. It is said that Handel wrote *Messiah* in just 24 days. He went into a trance-like flow, neglecting food and sleep to pour out the music. His mood verged on mania in this time; 'I thought I saw the face of God', was his comment when his maid found him in tears on the floor. In fact, he had created one of the supreme masterpieces of Western culture.

Kay Jamison's study of living artists and writers revealed that most of them had 'high' periods which they used to further their work. These mood changes tend to come on just before a productive episode and have many of the symptoms of an affective disturbance, such as early waking, restlessness, libido changes, and an increase either in enthusiasm and energy or in anxiety and fearfulness, or both. These moods will often be channelled into work output. As one person put it, 'I have a fever to write, and throw myself energetically into new projects'. Another described the intense stream of creation that can be attained in these times: 'work will flow almost as though one is a medium, rather than an originator'. Eight or nine per cent of the creative writers and artists Jamison studied reported mood episodes of this kind, and 90 per cent felt them to be necessary or very important to their work. What is more, they are right, since Jamison has shown that surges in productivity follow surges in mood, and that, like mood, productivity tends to be higher in the light months of the spring and summer than those of the winter.

Hypomania seems to be important for creative output for three reasons. First, it facilitates the speed and range of the imagination, so important for creative originality. The American poet Robert Lowell, during his manic attacks, wrote and revised furiously 'and with a kind of crooked brilliance', while talking about himself in connection with Achilles, Alexander, Hart Crane, Hitler, and Christ. Lowell describes his state of mind during one such attack:

> The night before I was locked up I ran about the streets of Bloomington crying out against devils and homosexuals. I believed I could stop cars and paralyze their forces by merely standing in the middle of the highway with my arms outspread . . . Bloomington stood for Joyce's hero and Christian regeneration, Indiana stood for the evil, unexorcised, aboriginal Indians. I suspected I was a reincarnation of the Holy Ghost . . .

The imaginative ranging was even more vivid for the manic-depressive poet Theodore Roethke:

Suddenly I knew how to enter into the life of everything around me. I knew how it felt to be a tree, a blade of grass, even a rabbit... One day I was passing a diner and all of a sudden I knew what it felt like to be a lion. I went into the diner and said to the counter-man, 'Bring me a steak. Don't cook it. Just bring it.' So he brought me this raw steak and I started eating it.

The second feature of hypomania that aids creative output is that it provides enormous energy to drive on through a task even in the absence of immediate rewards. Writing, painting, and composing are lonely occupations, which lack the online feedback that in other domains, such as sport and social interaction, keeps us motivated, concentrated, and happy. Any feedback that comes, and mostly it won't, will come years later when the person is working on something quite different. To ride through a difficult and enervating task, week in, week out, quite alone, without any validation from the outside world, one has to sustain an unreasonably enthusiastic mood. In fact, one has to be in a mood which, from the point of view of most other activities in life, is pathological. One should not blast on with unabashed cheerfulness in a relationship that gives nothing back for months, or persevere in an economic activity that seems to be yielding nothing. The adjustive function of the mood system should draw us gently away from these things. But the creator of imaginative products has to remain abnormally, almost irrationally, buoyant, and, to be successful, he has to produce a lot. Psychologist Dean Keith Simonton has shown that what distinguishes the most eminent producers from the rest in any cultural field is not that their work is consistently excellent. It is mainly that they produce a lot, and the more they produce, the more likely it is that some of it will be excellent. Schumann's hypomania helped him here, as did Mozart's and Handel's, and for that matter Picasso's, which allowed him to produce 14 000 paintings in his lifetime.

This relates to the third function of hypomania in creativity. We have already seen that depressives have a more pessimistic assessment of what they can achieve, but that, objectively speaking, that assessment is more accurate than that made by more sanguine individuals. That is to say, depressives are sadder but wiser, whereas most of the rest of us overrate our prospects. Now to take on a major imaginative project requires remarkable chutzpah. If, dear reader, you aspire to be a writer, poet, actor, artist, film director, or musician, then however you sell it to yourself, you must believe something like the following statement to be true. You have to believe that you can do something that is difficult, in a way that has never been done before, which will be of so much interest to your fellow creatures that they will reward you for it. But I have news for you, I am afraid. You are almost certainly wrong. I say this purely on statistical grounds. The vast majority of would-be writers, artists, musicians, and actors never become known for anything. This doesn't mean they don't have a good time trying, but it does mean that they were probably, in some orthodox sense, deeply unwise to follow the path that they did. This is the kind of benign folly that hypomania seems able to induce. It was Schumann, appropriately enough, who best summed up the endless creative optimism of his high moods:

> The goal we have attained is no longer a goal, and we yearn, and strive, and aim ever higher and higher, until the eyes close in death, and the storm-tossed body and soul lie slumbering in the grave.

I have argued, then, that the positive side of thymotypy, the side that aids creativity, is the ability to induce a sustained high mood. However, there is a problem with this account. Much the most common clinical symptom of the thymotypic genotype is not mania, but depression. In the population as viewed through the prism of medicine, the downs greatly exceed the ups, since only a small fraction of the people treated each year for depression have ever known the manic or even the hypomanic state. Depression is absolutely disastrous for creativity, as it slows the imagination, decreases the resolve, and lowers the sights. If depression is the most common effect of thymotypy, then surely the net effect of the genotype must be negative, not positive.

This may be a misleading way of looking at things. Thymotypy, as I argued in Chapter 4, is the liability of having one's moods spiral off from centre. When they do so, they acquire their own momentum, and become detached from immediate external signals. This comes to the attention of the clinician mainly when the spiral is downwards, or occasionally when it flies so high that it becomes dangerous. Mostly, the mild highs go unnoticed. It is true that there are millions of chronically depressive sufferers who have known little in the way of recompense for the drawbacks of their constitution. Their lifelong battle is with gloom, lethargy, and anxiety. However, for all I know, and I am speculating here, they have millions of cousins who are equally thymotypic, but have never had a psychiatric problem, because they have managed their lives in such a way that the spiral has always gone up, but not too far. Many others only ever report mood disorder in one direction—down, not because there are no ups, but because the ups appear to them not as disorder but as life richly lived. Creative writers have down periods when they produce little or nothing, but these are compensated for by the bursts of sunshine. The genotype has to allow both possibilities, since that is the way the mood system is built. The thymotypic genotype takes the brakes off the mood a little. It cannot predict which way the mood will roll when it does so. I can only assume that over evolutionary time, the ups have balanced the downs.

Schizotypy enhances creation in a quite different way. As we saw in the previous chapter, the key attribute of schizotypy as far as creativity is concerned is the capacity for divergent thinking. This fosters very significant originality in imaginative constructs, as studies using the Rorschach test and other materials showed. As an additional example, J. A. Keefe and P. A. Magano studied schizophrenics using a task where subjects had to come up with novel uses for a mundane object. They were scored for the total number of solutions, and for the originality of those solutions. Scores ranged from 0 or 1, for something irrelevant or stereotypical; to 2, for a rare use (e.g. for a pair of shoes, tie the laces together and hurl it like an Argentinean bolo); to 3, for something with superior imaginative content that goes beyond the object given (e.g. for eyeglasses, take the lens, fill it with water, and use it as a watering dish for your parakeet). Schizophrenics scored more highly on this task than controls, though, significantly, the subjects were young, recently hospitalized, and free from the worst paranoid symptoms at the time of the study.

Interestingly, it has been shown that healthy individuals in creative professions have an overlapping profile with schizophrenics. It is not just that they score highly on tests like the novel-uses task, although they do. Their underlying cognitive make-up seems similar. Like schizophrenics, they sample a wider range of stimuli than normals when there are several signals being fed to them. Thus some material comes into consciousness from a distractor that they are supposed to ignore. They have also been shown to pattern like schizophrenics, in some rather specific respects but not others, on personality tests. The Minnesota Multiphasic Personality Inventory is a battery of numerous personality scales designed to measure the propensity to various clinical problems. Although not schizophrenic themselves, a sample of artists and architects scored highly on the schizophrenia scale. Same capacity to take leaps of the mind, different outcome.

Theseus' description of the poet's imagination, from *A Midsummer Night's Dream*, reads remarkably like a description of schizotypy, the trait that (as we have seen) unites creative originality and liability to madness. Shakespeare's intuition would thus seem confirmed:

> The poet's eye, in a fine frenzy rolling,
> Doth glance from heaven to earth, from earth to heaven,
> And as imagination bodies forth

SUCH TRICKS HATH STRONG IMAGINATION 157

> The forms of things unknown, the poet's pen
> Turns them into shapes, and gives to airy nothing
> A local habitation and a name.

We have, then, a dimension of personality—psychoticism—which aids creativity. I have divided this into two, partly independent, subdimensions: thymotypy, which works mainly through the mood, and schizotypy, which works mainly through divergent thought. These two parts aid creativity in different ways. Perhaps the most potent force is some combination of the two, providing the drive and energy of hypomania, and the original thought of schizotypy. This is the pattern that emerges from biographical analyses of many of the most successful creators.

These temperaments, and the genes that underlie them, are deeply double-edged; the qualities that make them a blessing are the very same ones that make them a terrible curse. Balance is all. Over evolutionary time, we have to assume that the positive has balanced the negative, but at a human level, it is a lottery whether balance can be achieved. Because of psychoticism, many of our fellows who seem most blessed actually walk along a terrible knife-edge.

We should, once again, pause to take in the differences between the two conditions, as well as the parallels. The most severe form of thymotypic disorder is bipolar affective psychosis. Even this, terrible though it is, is essentially a transitory and reversible problem. For this reason, since way back in the nineteenth century, doctors have referred to it as *la folie circulaire*, or circular insanity, since cycling is the essence of the phenomenon. As long as there is a cyclical movement, there is hope. Between the manias and the depressions, there may be long periods of milder moods in which the person may function successfully. Sociological studies have shown, more often than not, that manic-depressives actually rise in the social hierarchy over their lifetimes. They end up disproportionately concentrated in the professional and managerial classes. Many of them, as we have seen, become eminent creators. They are able to do this because the thymotypic genotype, like any genotype, opens up several different roads, some good, some bad, but its particular genius is that it allows, nay forces, the person to sample several of them, hopping between them as he cycles.

This is in contrast to the schizotypic genotype. Here again, there are several roads opened up by the disposition. The choice at the junction is less reversible, though. Once a person has had a schizophrenic break, studies in most societies show that the chance of personal, social, and economic recovery is less encouraging. Schizophrenics tend to drift down the social hierarchy, ending up, as often as not, unable to provide for themselves. Kraepelin, once again, hit the mark with his choice of terminology; *dementia praecox*, not an image of cyclicity, but one of an irreversible and premature decline of the mind. This perhaps overstates the case; sizeable numbers of schizophrenics do return to completely normal functioning, but the majority do not, at least in Western societies. This does not mean that there are no schizotypes amongst the eminent. It is just that those who avoid the psychosis are the ones who, mainly, reap the benefits of the genotype. Heston's twin studies make sense in this light; the ill twin was very ill, and had no positive experiences, whereas the well twin was very successful and creative. In bipolar twins, you would be more likely to find that both twins had a mixture of the negative and positive features. Thus I would conclude that for thymotypy, the costs and benefits of the genotype are often both felt by the same person, whereas for schizotypy, the glory usually goes to one brother, and the pain to the other.

◆

The evidence reviewed in this chapter lends scientific weight to Shakespeare's premise. There is indeed a common root in madness and creativity. It resides in a type of mental temperament—strong imagination—and it is partly determined by genes. We should thus thank Shakespeare for his canny foresight, which led us eventually to the answer to our fundamental question, why there is madness in the human species? There is madness in the human species because it is an extreme form of normal human cognition, and the extreme forms have stayed around in the population because they produce both good and bad outcomes. The good outcome is unusual creativity, the bad one psychosis. The sword is double-edged.

However, the story is not finished here. Until a few thousand years ago, which is a mere blink of the eye for our species, there was no writing and there were no books. Hence there were no writers. There was certainly no musical recording. Whatever the ancestors of the Tennysons and Schumanns were doing, they were not attending literary salons or musical academies. Through most of history, people were hunter–gatherers. From what we know of those hunter–gatherer societies that have survived into the present, we can assume that not enough economic surplus was generated to support specialized professionals of any kind, least of all creative professionals. Everyone was focused on the more mundane task of staying alive.

This is a problem for the creativity-benefit argument I have put forward in this chapter. All the psychosis-prone creators I have cited in this chapter come from the bourgeois sector of Western societies that had achieved economic take-off. The creativity benefit they and their kind enjoyed may well be a life advantage now, in the warm luxury of the agricultural, industrial age. Nowadays, the basic needs of society can be provided by a few highly efficient producers. This frees the rest of us for more arcane pursuits, which we revel in perhaps precisely *because* we have solved some of our more pragmatic difficulties. But unless creativity was also an advantage back when people lived from hand to mouth, then the creativity–benefit argument must fail. This is because the human gene pool changes only very slowly. There has not been nearly enough time for it to come to reflect the conditions of contemporary life. It must reflect the adaptive situation of our hunter–gatherer ancestors. For the creativity–benefit argument to work, creative achievement must have been just as valuable to them as to us. It is not obvious why it should have been: you cannot eat a poem, scare predators with an old show tune, or keep warm by huddling round a cave painting. This issue must be addressed. If it turned out that creativity was not important to all societies, then the creativity–benefit argument must be rethought. If it turned out that it *was* important to all societies, then that would say something very interesting about what it is like to be human. It is to these matters that I now turn.

CHAPTER 7

The lunatic, the lover, and the poet

> I am, indeed, every day of my yet spared life, more and more grateful that my mind is capable of imaginative vision, and liable to the noble dangers of delusion which separate the speculative intellect of humanity from the dreamless instinct of brutes.
>
> John Ruskin

Orpingalik's most famous song, regularly requested in dance houses all over the country, was the one called *My Breath*. The song's first lines explain the choice of title:

> My breath.
> This is what I call this song, for it is just as
> Necessary to me to sing as it is to breathe.

Song was the very essence of Orpingalik's identity. Not only was he famed for it—he was perhaps the leading artiste of his generation—but it served an important psychological function for him. It was his comrade in solitude, and his natural release in times of suffering, as he explained eloquently:

> How many songs I have I cannot tell you. I keep no count of such things. There are so many occasions in one's life when a joy or a sorrow is felt in such a way that the desire comes to sing; and so I only know that I have many songs. All my being is song, and I sing as I draw breath.

Orpingalik's oeuvre has never been properly catalogued. He never had a recording contract nor cut a disc. In fact he never received any money for his performances. This is not because they were not valued. It is because no money ever passed through his hands at all.

Orpingalik was an Inuk. This means he belonged to the Inuit, the people of the far North who Europeans used to call, wrongly, the Eskimos. He lived in a traditional hunting and fishing society in the frozen wastes of northern Canada in the first decades of this century. It is only by an accident that anything of his life is written down. He happened to impress the chief ethnographer of a Danish expedition to the far North between 1921 and 1924, and this ethnographer recorded some details in his report.

Inuit society is a fluid affair, the basic social unit of which is the nuclear family. The country is bounteous in many ways, yielding to the skilled hunter a crop of seals, salmon, caribou, musk ox, ptarmigan, wolves, wolverines, and bears. However, conditions are desperately harsh, with temperatures rarely exceeding freezing point and often far, far lower. Families must move around their territory with the seasons, residing in snow huts or tents, following the game and fish. Dogs and even babies that cannot be supported must be unsentimentally disposed of. Scarcely any possessions can be amassed, save the most practical implements of survival—furs, hides, hunting and fishing equipment—since everything must be either carried or left about the country in rock-top caches.

Although theirs is a hard and empty world, Inuit families are not isolated. They join up in bands of varying sizes at different seasons, and these bands drift together or apart as they move around. The centre of any band that forms is the dance house. This is an enlarged tent or snow hut in which the people can gather to sing and dance. When two Inuit groups visit each other, a formal dance, in which leading individuals from both parties take their turn to do their stuff, is the first priority. The dances, though social, are not collective endeavours. Dancers take the arena, one at a time, and produce a song or poem, accompanied by a slow dance and a small drum. The songs treat of events that concern the singers, fortune in hunting, relationships, life events, and the natural world. Although there are recognizable patterns and standards, both in the music and the words, the essence of the art is not to preserve tradition, but to provide new variations, permutations, and innovations on the old themes, thus stretching the genre in new directions. The effect is impromptu, but this does not mean that the songs are made up on the spur of the moment. In fact, considerable prior thought goes into composition. The creative process itself is the subject of many songs, as when Piuvkaq sings of standing at a hole in the ice, jigging for trout and seeking inspiration:

> Why, I wonder . . .
> My song-to-be that I wish to use
> My song-to-be that I wish to put together
> I wonder why it will not come to me.

There are no professional dance-singers, in the sense of people who make performance or composition their main business in life, but then, there are no specialized professions amongst the Inuit. Everyone is, must be, a hunter, a fisher, and builder of snow huts, as well as anything else. This does not mean that dance-singing is a marginal activity, as the report of another ethnographic expedition, in 1914–16, noted:

> Every Eskimo, therefore, whether man or woman, can not only sing and dance, but can even in some measure compose dance-songs. Distinction in this field ranks almost as high as distinction in hunting, for the man who can improvise an appropriate song for any special occasion, or at least adapt new words to an old song, is a very valuable adjunct to the community. Certain individuals naturally possess greater ability than others; their songs become the most popular and spread far and wide.

Orpingalik was just such a man. As well as an expert dance-singer, he was a shaman. Shamans are religious figures found throughout the societies of the far North. They hold séances during which they become inhabited by the spirits of animals or the dead. Their utterances during these trances are oracular, and to be taken seriously as warnings of the future. Shamans have tremendous magical powers. Their office, though, is neither full-time nor monopolistic. That is to say, they must be hunters and everything else as well as shamans, and, moreover, there are many competing shamans in any group. There is no nice shamanic parish anywhere with a cosy vicarage attached. Instead, shamans constantly vie for influence through the quality of their séances and the numinousness of their visions. Shamanic power and skill in song are not entirely separable; as in medieval Europe and so many other societies, the religious and secular domains of creativity are intimately intertwined.

The Inuit example is useful as an illustration of several key points. The first is that creativity is not just valued in modern, affluent societies. The Inuit, in their treacherous environment, are preoccupied with the needs of subsistence as much as any society ever has been. In fact, they are really up against it. They have almost no economic surplus, and, with no permanent villages, kings, or chiefs, their social organization is about as basic as can be imagined. Indeed, the form of it probably resembles that of early hunter–gather societies the world over, a form which, in gentler climes, has been gradually superseded by something more large-scale.

What can be said to be central to the life of an Inuit band? Hunting, sure; fishing, undoubtedly; family-formation and child-rearing, beyond question. But there, at the very heart of the settlement, stands the dance house, and all that it entails. The first thing people do on coming together is show off their dance-songs, and skill in this 'ranks almost as high as distinction in hunting'. Yet these songs, though no doubt enjoyable, seem to lack any obvious survival function. They are fictions and reflections, and fictions and reflections cannot keep out the icy wind, or lure a bear into the cooking pot.

The Inuit example, although especially neat, is not at all special. In all human societies that have ever been studied, there exist forms of skilled performance that lack any obvious practical value. Of course, the form and content of these performances vary enormously over time and between cultures. They can be visual, verbal, or musical, but in some way or other they always exist. The essence of good performance is always artful innovation on a recognized theme, stretching and expanding a genre by putting a personal stamp on it. The sociology of performance is also remarkably constant. Skill in it is highly variable amongst the population, with many people trying out, and just a few individuals achieving great eminence. As soon as a society produces enough surplus to support non-producing castes, such individuals can live by their art. This can be observed from the bards of early Europe to the singers of praise-songs and epics in the agricultural societies of West Africa. Elsewhere, creators must toil like everyone else, and exercise their *métier* in spare time for the accumulation of prestige alone.

The universality of creative performance has very important implications. For one thing, it suggests that there would be a positive outlet for the traits of psychoticism in all populations. Schizotypy is strongly suggested, for example, by the shaman's capacity to hear voices, to have paranormal visions, and to be inhabited by imaginary spirits. The thymotype's capacity for hypomania would be as useful to a bard in a traditional society as it was to Schumann. Indeed, there are ethnographic examples of thymotypic traits in traditional creators. The Fijian oral poet and seer Velema, whose oeuvre was studied by anthropologists in the 1940s, was widely reknowned and became a 'big man' on his island. From early in his life he showed a diffident and excitable personality. As he aged, he became more and more eccentric, often brooding all day in the house or wandering alone in the forest. These behaviours, which suggest affective disorder and would normally be reprehensible in Fijian society, were felt to be compensated for by Velema's visionary work.

Even our Inuk friend Orpingalik might have had a thymotypic disposition. His masterpiece *My Breath* is about a period of unspecified illness he suffered. We don't know what this illness was, but the song calls very much to mind the lethargy, fear, and hopelessness of depression. *My Breath*, despite coming from a vastly different culture thousands of miles away, is instantly and strikingly accessible, especially to anyone who listens to blues music:

> I will sing a song.
> A song that is strong,
> > Unaya—unaya.
> Sick I have lain since autumn,
> Helpless lay I, as were I
> My own child.
>
> Sad, I would that my woman
> Were away to another house
> To a husband
> Who can be her refuge,
> Safe and secure as winter ice.
> > Unaya—unaya.
>
> Sad, I would that my woman
> Were gone to a better protector
> Now that I lack strength
> To rise from my couch.
> > Unaya—unaya.
>
> Dost thou know thyself?
> So little thou knowest of thyself.
> Feeble I lie here on my bench
> And only my memories are strong!
> > Unaya—unaya.
>
> Beasts of the hunt! Big game!
> Oft the fleeing quarry I chased!
> Let me live again and remember,
> Forgetting my weakness.
> > Unaya—unaya.
>
> That was the manner of me then.
> Now I lie feeble on my bench
> Unable even a little blubber to get
> For my wife's stone lamp.

The time, the time will not pass,
While dawn gives place to dawn
And spring is upon the village.
   Unaya—unaya.
But how long will I lie here?
How long?
And how long must she go a-begging
For fat for her lamp.
For skins for clothing
And food for a meal?
A helpless thing—a defenceless woman.
   Unaya—unaya.
Knowest thou thyself?
So little thou knowest of thyself!
While dawn gives place to dawn,
And spring is upon the village.
   Unaya—unaya.

 The familiar picture from the last chapter, in which psychoticism has adaptive value through creativity as well as negative effects through psychosis, is thus confirmed for this hunter–gatherer society as well. The argument that creativity is the benefit keeping psychotic alleles in the gene pool stands firm. Note, too, that the Inuit example corroborates the finding from Western societies; the valued output of people with high psychoticism is not practical forms of creativity, like devising better hunting techniques, it is creativity of a purely imaginative kind, the making of representations which are complex, aesthetic, individualized, and which have no use value beyond themselves and whatever thoughts they strike up in their audience. This confirms the pattern seen in the studies of the biographies of eminent people in our own culture. The big excess of psychosis is not in practically creative domains, such as engineering, science, and architecture, but in aesthetically creative ones, such as poetry, visual art, and music.
 Aesthetic creativity, then, is not just an add-on, which societies do when they find themselves with spare time and

energy. Rather, it is central to humanity in general, and so valued that individuals with an advantage in it flourish despite the costs inherent in their personalities. But why should this be so? It strikes me as a deep paradox about human nature. Anthropologists have generally assumed that our powerful cognitive capacity evolved to enable us to handle sophisticated, useful truths about the world; tools, hunting, solving social problems, and so on. Yet this is not what fills up much of the mind. Most books that are written, most songs that are sung, most paintings that are painted, are fictions. The things that they assert about the world are, at a literal level, falsehoods. Yet these are the things we know and remember, and discuss incessantly. Fiction outsells non-fiction the world over, and cultural prestige goes not to the achievers of practical ends but to the spinners of good yarns. You remember Chaucer and Shakespeare, but can you summon the name of the person who invented the vertical-sail windmill, or the shoulder-seated horse collar? These latter things revolutionized European life, since they enormously enhanced the efficiency of food production, but their originators are lost in the mists of time. You remember Beethoven and Brahms, but can you name a single innovator in the field of sewer construction and sewage treatment? You cannot, and indeed the idea is laughable, but it shouldn't be. Creators in the field of sewage treatment have saved hundreds of millions of lives in the past two hundred years. They have done as much for humanity, in terms of living and dying, as anyone else, yet to say that they summed up the spirit of their age would be ludicrous. On the other hand, it is quite common to say this about a musician or a painter, whose effect on ordinary life is much more nebulous. Our choice of culture heroes seems wilfully impractical.

Questions, questions. Our starting question—why should there be madness in the human species?—has found an answer—because the traits that underlie madness are also beneficial for creativity—but this only lets the deeper question flood in. Why should creativity be so advantageous? And why should our cultures be so obsessed with it?

The peacock's tail is, quite frankly, an affront to evolution. The theory of natural selection implies that animals will evolve to be desperately utilitarian; the right tools to gather food, the right camouflage to avoid predators, everything as economical and efficient as possible. And indeed, the peahen, in her drab plumage, does nothing to upset this picture. But the peacock flouts it with gusto. His enormous, gaudy, shimmering tail is longer than his body. When he opens it, the dazzling greens set off the bright blue of his torso, and to cap it all, a row of huge, Daliesque eyes peer out at his audience. It is a work of art.

Darwin knew that the existence of the peacock's tail was a problem for his theory of natural selection, as first formulated. 'The sight of a feather in a peacock's tail, whenever I gaze at it', he wrote, 'makes me sick'. He hinted at a possible solution to the problem in *The Origin of the Species*, and returned to the problem in detail in his later book *The Descent of Man, and Selection in Relation to Sex*. The problem was, quite simply, this: all over the natural world, there are structures and behaviours that seem wilfully, if not dangerously, impractical. The peacock's tail is an extreme example, but we could equally cite the colouration of many fish, bird, and insect species, the bowers of the bower bird, the crests of cockerels, or the bright face of the mandrill. These things are expensive to grow, and make the owner, at the very least, more conspicuous, and quite possibly less able to get on with life. Indeed, there is direct evidence of their impracticality. The scarlet-tufted malachite is a Kenyan bird species, the males of which have a pair of long streamers attached to their tails. When scientists artificially shortened these streamers, the birds became better at catching insects, and when they lengthened them, they became worse. The streamers are an encumbrance, and yet they persist. Why could this be?

Darwin's answer was to recognize that evolution is not really about the struggle to survive so much as the competition to reproduce. If the males with the long streamers were much better at attracting mates than those without, then as long as they could stay alive long enough to mate, they would predominate, despite their ludicrous addenda. If females preferred ornamented males, then ornamentation could evolve, even if its costs were significant. This, then, is the answer: males have long streamers in order to seduce, and being able to seduce is as important as being able to eat in terms of evolutionary outcomes. This is the central idea of the theory of sexual selection, or, more exactly, the theory of intersexual selection, which deals with mate choice and its effect on evolution.

The idea that choice of partner by one sex affected the evolution of the other was neglected for nearly one hundred years after Darwin. For one thing, it was too threatening to admit the caprices of fashion and the aesthetic fancies of females as central motors in the history of life on Earth. Other, more workaday explanations for gaudy colouring were sought, such as use in same-sex conflicts and signalling to predators. However, the topic has returned to prominence over the past twenty years, and, using advanced mathematical models, the basic idea has been vindicated. Sexual selection is now recognized as a central component of the evolutionary process.

Exactly why it happens is still a matter of debate. Male scarlet-tufted malachites with long streamers do indeed attract more females than those with shorter ones, so there is no doubt that female preference keeps this trait alive. But why should the females have such a preference? This is the difficult question.

An early answer was provided by Sir Ronald Fisher. The preference just *is*, and it cannot stop being because of a kind of evolutionary tyranny of fashion. Imagine a population in which males with slightly longer tales attract more mates. Females who mate with long-tailed males will have sons who also have long tails. In turn, those sons will get more mates, so in terms of grandchildren, the females with the preference for the longer tales will fare better. Any male without a long tail suffers a disadvantage because he is not as attractive as his rivals. More crucially, any female who does not prefer long tails suffers a disadvantage because her sons will not be as attractive as their rivals. So neither male nor female can break the cycle, and the tail and the preference must co-evolve inexorably until everyone's tail is as long as it feasibly can be.

Now there are two problems with this theory (which is known as the 'run-away' model, since an initially arbitrary preference leads to great effects). The first is that, sooner or later, everyone ends up with the long tail. This means that it no longer matters who the females choose. Their sons will be equally attractive in any case. Thus the selection pressure keeping the female preference in the population disappears. This means that although the run-away process can produce a period of rapid change in traits under sexual selection, it may not be sufficient to account for their maintenance in the long term. The second problem is that the model assumes an intrinsic preference for the long tail in the first place. This is required to get the process started. This seems like a circularity, though in fact it may be fairly innocuous. For example, long things are just more striking to the senses, as are bright things and loud things. Thus the initial effect may be as simple as a small perceptual bias on the part of the females to whatever makes the greatest splash. The peacock's tail may have a row of eyes in it since eyes are something that peahens are pre-adapted to notice. The male is exploiting the way the female's psyche happens to be set up. The question is whether this kind of perceptual bias effect is enough to explain the very elaborate form that sexual ornaments take.

The main alternative to the run-away model is known as the good-genes theory. This theory stresses that female preferences are not arbitrary, but are rather discriminating. The female wants to make sure that the 50 per cent of foreign genes she is introducing into her offspring are the best she can possibly get. She thus looks for traits that are reliable indicators of male quality, resilience, health, and so on. Under the good-genes theory, as developed by Israeli biologist Amotz Zahavi in particular, the traits females should be most interested in are precisely those that are most costly to bear. If it is difficult and impractical to produce and maintain a huge, gaudy, symmetrical, patterned tail, then any male which does so and remains alive must be in pretty good shape. He must be relatively free of disease, good at procuring food, and canny at evading predators, and these are all the things the female wants for her children. It thus pays her to direct her attention to the difficult, impractical traits that only a few individuals can pull off, and it pays the males, as soon as they have any energy surplus from the business of surviving, to invest it in just such things.

In the good-genes model, then, the form of sexually selected traits is deliberately outlandish, because it is the really useless, showy, and costly structures that best indicate underlying quality. It is as if the male is saying, 'look at me, I can do all this and still manage to do all the normal business of surviving'. This is known as the handicap principle, and the *folie de grandeur* that is the peacock's tail seems to be nicely explained by it. Perceptual bias effects are not ruled out in the good-genes theory, though, since of two tails that are equally costly, the one that most grabs the female's attention will obviously do its bearer most good. What evolves, then, are highly costly, impractical traits, which are relatively good at striking the senses of the female.

Most biologists believe that sexually selected traits evolve by female choice, through some combination of run-away and good-genes effects. This process never ends, because the ever-changing micro-environment, not least of diseases and parasites, means that the genes that are good for one generation are not good for the next. Males must constantly display their quality, and females must be ever choosy, for sharing your

THE LUNATIC, THE LOVER, AND THE POET 173

genes with somebody is a mighty decision. Note that the sex division of roles in sexual selection need not always be males displaying and females choosing. It can work both ways. In nature as a whole, it most often goes that way around because females, by and large, have much more at stake in any single copulation than do males. This is because a female will be stuck with the costs of the offspring; the male may just disappear afterwards. He can have many more fertilizations in his life than she can have ovulations, and so she should be more guarded, and he more urgent. Humans are not completely typical of this pattern, since, although most societies are polygamous, men invest in the raising of their children in a way the average male mammal does not.

How might one apply the lessons of sexual selection theory to the current problem? Certainly, there is an obvious similarity between the peacock's tail and what goes on in the Inuit dance house. There, at the very centre of the struggle to survive, one suddenly encounters a thing of deep impracticality and showiness, in which individuals compete to impress each other with their useless but highly ornamented structures, structures they have spent their spare energy and time producing, and for which a few of them will reap great prestige. Human creative performance could well be, at root, a form of sexual display.

This is the theory that has been put forward by the evolutionary psychologist Geoffrey Miller. It has a compelling appeal to it, but proving it is a little more difficult, and so we must ask what actual evidence we can bring to bear on the question. If the theory is true, then, other things being equal, more creative individuals should have higher mating success than less creative ones. Anecdotally, this is undeniable. The king of the Afrobeat sound, Fela Kuti, once married 17 of his dancers in a single night, and it is never difficult to find examples of the sexual hysteria that surrounds particular rock stars, writers, and artists. Statistically, the relationship between artistic and sexual success has not been so well demonstrated, and indeed it is hard to do so, because of the 'other things being equal' condition. In modern societies, the criteria of mate choice are very variable across space, time, and social class, so the relevant comparisons are hard to make. Not everyone plays by the same rules. Mating success relates to money in some sectors of society, and other factors like intelligence, beauty, or education in others, in such a way that generalizations are difficult.

However, the theory does not need to demonstrate that creative people have greater mating success now, in our developed and allegedly monogamous societies, which are so different from the societies in which we evolved. Rather, the theory must show that the *ancestors* of creative people enjoyed such an advantage in the societies in which they lived. This is hard to do directly, for obvious reasons. Ethnographic studies of traditional societies show one thing very clearly: there is a general correlation between economic and reproductive success. This is especially marked in the majority of traditional societies which are polygamous, since a man's economic capital determines how many wives he can manage to take, as well as the survival rate amongst his children.

Ethnographic studies also show another effect: the existence, in all societies, of less concretely economic resources, which are also tied up with social and reproductive success. These are often referred to, following the French sociologist Pierre Bourdieu, as cultural or symbolic capital. Skill in the dance house would be an Inuit example. In other societies, it might be painting or the ability to communicate with the spirit world. These types of capital are alternative routes to reproductive success and to social reknown. Such routes exist very obviously in our own culture as well, as the image of the impecunious poet testifies. The enduring glamour of professions such as poetry, acting, and music, which, as many a hopeful has discovered, is more in the idea than in the reality, has the heady scent of sex about it.

Pursuing cultural capital, then, is one way of increasing mating success, and the main way to do this, across cultures, is through the appropriate kind of creative performance. This observation supports Miller's theory that creativity is a sexually selected trait. Further testament to it is the fact that we are so interested in creative people. They are the most famous human beings there are, and our appetite for learning about them is limitless. The next best-selling category of book after fiction is biography, and biographies are written not just about the rich and powerful (the holders of economic capital), but also about the creative (the holders of cultural capital).

Such generalizations as these are obviously not conclusive. However, there are ancillary lines of evidence that Miller brings to bear in support of his theory. If creativity were a sexually selected trait, then it would share some hallmarks of other such traits. What are these hallmarks?

Firstly, there is abundant variation between individuals of the same species in sexually selected traits. This variation is used by the other sex in making their mate choices. By contrast, traits shaped purely by survival factors tend to homogenize on an optimal, utilitarian norm, with everyone showing them to the same degree. This is the individuals effect. Secondly, birds do not put on their gaudy plumage until puberty, and often only keep it on for the mating season. This is the time effect: sexually selected traits are most visible at the time of maximum reproductive opportunity. Thirdly, the sex that has most to gain from additional matings invests more heavily in sexual display traits than the other. That is why the peacock is more splendid than the peahen. This is the sex effect: the sex that invests least in each mating has the traits more visibly than the other sex. Does human creativity show any of these hallmarks?

The social profile of creative output is indeed consistent with the idea that its original function was sexual display. There is enormous variation between individuals in creative output, too much for creative output to have been absolutely necessary for survival in our ancestors' environments. Miller has also demonstrated interesting age–sex effects, at least for modern Western society. Output, whether of music, books, or paintings, rises dramatically at the age of sexual maturity. The age of maximum output varies a little according to the medium, presumably depending on the time required to learn the craft, but it peaks in the prime of reproductive life, and declines thereafter. (This pattern, the reader may recall, mirrors that for the onset of psychosis.)

There is also a sex effect. Across all media, men produce about ten times more cultural performance than women. This is hard to interpret. Evolutionary psychologists would argue that this represents an evolved, biological difference between men and women in the drive to cultural performance. The traditional social science explanation, by contrast, would be that the gender imbalance is a consequence of the socio-economic roles women happen to have been assigned in this

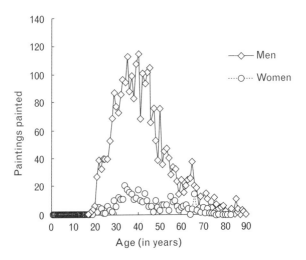

Figure 10 Output of modern paintings as a function of age and sex of the painter. (From Miller, G. (1999) Sexual selection for cultural displays. In *The Evolution of Culture*, (ed. C. Knight, C. Power, and R. I. M. Dunbar). Reprinted with the permission of Edinburgh University Press.)

period of history. There is no doubt that the social science explanation is in some sense correct in this case. We are a supremely flexible species, who apply our numerous behavioural strategies in a way that is highly sensitive to local context. We are not slaves to any simple instincts, as the evolutionary explanation would seem to imply. We can thus look forward to a time when sex roles in culture and economy may be different. However, this said, there is a germ of truth in the evolutionary account, too. For most of history, the variation in men's reproductive success has been much greater than that in women's, so it does follow quite naturally from sexual selection theory that men should have been more driven to cultural performance than women, and this is what Miller's data show.

Miller's theory, then, seems a highly promising explanation of cultural performance. It obviously awaits the amassing of direct, detailed cross-cultural evidence that intelligence and creativity are indeed key criteria of human mate choice. In the meantime, though, it makes more sense than any alternative approach. It even suggests one way that Shakespeare could have been onto something when he interposed 'the lover' between 'the lunatic' and 'the poet'.

However, I should be very clear about what Miller's theory does and does not mean. For one thing, it does not mean that the conscious or even unconscious motivation in the head of the creator is the desire to attract a mate. This has been claimed from time to time, not least by Freud, for whom creativity, as much else, was a sublimation of the sex drive. This is not the implication of the sexual selection theory. Miller merely claims that the reason that the drive to create has stayed around is that it makes the people who have it attractive to the opposite sex. Thus they leave descendants. What they are actually thinking about when they exercise their creative drive is quite another matter. For many creative people, creation is either an end in itself, or serves a very personal function in making their world make sense. They are not thinking about showing off, but Miller's theory does not require that they do.

Just as Miller's theory is irrelevant to what is going on in the head of the creator, it is largely irrelevant to what is going on in the head of the audience as well. He does not claim that music lovers are really thinking about possible matings with the singers when they enjoy an opera, though I must admit that such thoughts have struck me more than a few times while listening to the Habanera from Bizet's *Carmen*. They are thinking about the content of the show, which they value for itself. The sexual selection theory merely says why they should be disposed to value it, and therefore to be interested in the people that produce it. The theory is not, then, a philistine reduction of all aesthetics to a subconscious drive for sex, nor a denial of the power of culture to educate, charm, ennoble, shock, or entertain. It is a theory about the evolutionary significance of cultural performance, not its human significance.

That said, the theory does give us some insight into why cultural performance is about the things it is about, and what constitutes quality in it. As we saw in our discussion of sexual selection in general, a good sexual display is whatever, per calorie invested, produces the greatest effect on the senses of the audience. Animals with monochrome vision do not grow colourful tails, and snakes, which are deaf, do not sing as birds do. There would be no point. Instead, they must come up with a form of display that will weedle itself into the minds of others. It must thus chime with the existing mental set-up of the rest of the species. On the other hand, the whole point of the display is a competitive one. It must show the displayer to be fitter than anyone else around. There must therefore be an element of novelty or improvement. These are the twin desiderata of a sexual display: to do something new enough to make you memorable, but not so completely original or bizarre that your audience cannot recognize it.

This is precisely how culture works. Successful performances are good at striking our senses and intellects, and

THE LUNATIC, THE LOVER, AND THE POET 179

sticking there. Fiction and drama must depict situations that are intrinsically interesting to us, and resolve them in ways that stimulate or satisfy. Music must exploit the quirks of acoustic physics, and of our auditory brain, which was made to decode language. Thus musical styles must exploit the fundamental harmonies of nature, and also the basic phrasal cadences, rhythms, and stresses we find in speech. Visual art must hit our natural buttons of interest in symmetry, colour, human faces and bodies, and landscapes. It can hit these by pandering to them, or by disconcerting them. The best art does a bit of both.

Thus successful cultural creations must chime with what we already know and perceive about the world, connect with our prior experiences and expectations, and lodge themselves in our minds because they hook up with meanings we already had stored. On the other hand, anything that *only* does this is instantly forgettable, and will not be valued in the long run. Enduring art has an element of originality or innovation, of going beyond existing genres and traditions, of putting a personal stamp, of varying the theme, of throwing an inexhaustible and enduring bit of surprise into a general comprehensibility. Enduring culture fills the mental bucket of its audience, and overflows a bit. Mediocre culture merely fills the bucket. Avant-garde culture often misses the bucket altogether, and as a result remains a minority interest, suspected by many to be nothing more than a mess on the floor.

This is why the cultural canon is ever-changing. As the preconceptions and concerns of audiences change, so creative representations change, the better to capture them. New genres and styles are always being created, usually by breaking the mould of their predecessors bit by bit, and every serious creator wants to push the limits of his medium, to take it somewhere it hasn't entirely been before. We could not stop with the music of Handel, perfect as it might seem. From baroque music, classical had to come, from classical, romantic; from romantic, modernist and atonal music. No composer goes into his craft to leave it the same, so there was no real choice but to explore these new pastures. If the sexual selection theory is right, then this momentum in culture is unstoppable, because the whole point of culture, at the evolutionary level, is to demonstrate the imaginative ability to surpass all previous productions in both memorability and originality.

◆

Another issue is raised by Miller's sexual-selection theory. If humans had to have a domain of sexual display, then why culture? Although we have no data that could decide this question, it is not a vacuous one. In principle, any kind of costly, difficult to maintain structure or behaviour would do. Perhaps plumage became the principal sexual choice arena for birds because feathers are, in the end, crucially important to their lives. The choice of the domain of sexual display reflects this. The bright plumage takes a pre-existing domain of perceptual interest—feathers—and fictionalizes it, which is to say, it takes it away from the pragmatics of flight and makes it fabulous. Perhaps tropical fish are bright for the same reason: scales are central to fish life, and so fish would necessarily be interested in healthy scales. Bright colouration takes that interest and runs away with it. Can similar reasoning be applied to human culture?

It is very interesting that, if Miller is right, the central domain of human sexual display is not a physical characteristic, but a cerebral one. The skills that are directly demonstrated by cultural performance relate to perception, imagination, intelligence, and mental application, not, for example, strength, sexual prowess, or hunting ability. I believe this is no accident. The most striking single theme in human evolution is the relentless increase in brain size. This increase occurred despite the enormous energetic costs of building and running a brain as large as ours. The only explanation for this trend is that there was an enormous pay-off for being brainy.

The most promising current theory of what such a pay-off would be is that braininess was necessary for managing in the large and complex social groups in which we came to live. To live in a such a group, you have to be able to monitor the behaviour of others, keep track of who is doing what to whom, infer what will happen if you do something to someone, and plan how you will deal with conflicts between you and someone else. This requires a lot of computing power. It requires not just storage space, but also an engine of planning and inference of the sort which says, 'If I do *a* to *b*, then if *c* does *e* to *f*, *g* will think that *h* . . .'. It further requires skills of interpersonal perception, the ability to put oneself into another person's position, and the ability to influence people in novel ways. Once language had evolved, these requirements became much more intense, for there were now vast new channels for understanding, manipulating, and relating to each other.

I would contend, then, that one of the main criteria for mate choice in human evolution would have been whether or not the potential spouse had these mental abilities, because if he didn't, your children probably wouldn't either, and that would be a problem for them. Culture takes off from that point, by basing itself in the desired mental qualities and showing them off in superabundance. The major cultural forms are all deeply brainy. They exhibit rare skills of intelligence, mental application, interpersonal perception, the ability to put oneself in another's shoes, planning, the ability to anticipate the effects of something on someone else, and the ability to make inferences about the behaviour of others. Even musical composition, apparently the most abstract of media, is about envisaging and planning a structure that will have a certain effect on the emotions of others. And these elements are obviously central to song, poetry, theatre, fiction, and visual art. Thus the essence of cultural performance is taking mental skills and showing off an excess of them to such an extent that they go well beyond reality or any practical outcome.

What this amounts to is the claim that cultural performance is simply a way of demonstrating your fitness to build and run an exceptional brain. Miller's view is that sexual selection has acted on, and thus increased, the level of IQ in the human population. This is very possibly correct, but I believe that the evolution of psychoticism could also be explained by the sexual selection theory. Once sexual selection for intelligence had begun, and cultural performance was being used as a marker for this, then any mutations along other dimensions that enhanced cultural performance would also be adaptive. Psychoticism can do this, as we have seen. The schizotypy subdimension does so by enhancing originality of thought, and the thymotypy subdimension does so by enhancing the creative mood. Sexual selection would have thus maintained higher levels of these traits in the population than would be adaptive for more utilitarian reasons. The cost that prevents levels of psychoticism becoming even higher is the increased risk of psychosis at the top end of the distribution. This is why not everyone has high levels of the trait, for those that do suffer terrible costs as often as potential benefits. At the population level, we can only assume that these costs and benefits are in some kind of evolutionary balance.

If this account is fundamentally correct, it says a great deal about what we as a species are really like. Anthropologists and archaeologists have usually characterized the uniqueness of humanity in terms of cleverness and practicality. They dubbed us *Homo sapiens*, the wise man, thinking of our reasoning skills, or else *Homo faber*, man the maker of tools, thinking of the practical, concrete ends to which we put our intellects. Such designations put the practical achievements of humanity centre stage, and aesthetics to the margins. But if the latest period of human evolution has been driven to a significant degree by sexual selection for cultural performance, then we can think of ourselves slightly differently, in a way that chimes more with our subjective experience of what we are. We are *Homo imaginans*, the makers of imaginative, subjective, impractical, often fictional representations of our situation, which we use to charm, delight, and impress on each other. Madness exists because we have been selected to be a little removed from reality, to be concerned with the unreal. We have selected this way because it allows us to put on good cultural displays.

The idea of *Homo imaginans* is a powerful one, which could be used to understand some of the deep divisions that exist between scientific and humanistic ways of understanding ourselves. Such divisions are the cause of much misunderstanding in the human sciences. They have nearly destroyed the discipline in which I received much of my academic training, anthropology. Within anthropology there is a huge and acrimonious division between, crudely speaking, those who stress that societies are everywhere the same, and those who stress that societies are everywhere different.

In the former camp are people such as economists, evolutionary psychologists, and sociobiologists. They argue that human behaviour follows the same basic patterns everywhere. People have pursued the same fundamental goals of survival, optimal subsistence, mate choice, and reproduction, using the same problem-solving abilities and basic motivations, at all times and in all places. And intuitively, there is a lot to this position. Viewed in purely behavioural terms, human societies are pretty uniform. People are largely concerned with extracting resources from their environment in the most opportune way, whether with the fishing-spear, the digging stick, the plough, or the credit card. They maximize their economic and cultural capital as far as possible. They live in bonded social nuclei based around child-rearing, and they tend to produce children when resources and timing permit. They have systems of kinship which roughly reflect genetic relatedness, and they favour their kin over non-kin in certain ways. They form reciprocal relationships with non-kin, and the breakdown of these relationships is a major source of conflict for them, as are resource competition, status competition, and, above all, mate choice.

All these things are fairly banal and predictable to a hard-headed Darwinian. Any differences in the details of how different societies actually carry them out just reflect the fact that they live in different micro-environments. So we are, then, just another primate species, not doing anything terribly surprising.

But in another way, there *is* something deeply surprising about humans. This is where the opposite camp of anthropologists, the social and cultural anthropologists, have their say. Those on this side of the debate point out that what it is like to be a human, from the inside, varies a lot according to where and when you happen to be born. If you are an Inuk, your understanding of your universe—how it was created, what spirits it contains, how to make your way in it, what will happen to it in the future—will be completely different from that of a Christian. Although some behaviours, such as religious rituals, may look homologous across societies, the cultural logic of them is quite different from place to place. In one society they will exist because the ancestors must be appeased. In another, they function to summon the spirit of the bear, which is to be hunted. In another, they serve to placate the sun, so that he will weep tears of rain. In yet another, they serve to ensure happiness in the afterlife. In short, the behaviour is the same, but the story we tell ourselves about it is always different. It is the same with something like kinship; some societies say we should be nice to our brothers because blood is thicker than water; for others it is because we share the same totem animal, who would be angered if we did not; for others it is because the ancestors would bewitch us otherwise. Same behaviour; new, imaginative, different story.

Social and cultural anthropologists thus stress that what it is to be human is not universal, but is instead constituted by our culture. The strength of their approach is its ability to explore what the world of a particular society is like, to follow its logic, mythology, and ideology. In a way, the divide within anthropology is like the dichotomy between scientific and humanistic psychiatry all over again. One side of psychiatry concentrates on the universal biological mechanisms of psychic disturbance, the other on the personal universe of meanings that make up the human individual. More generally, there is a kind of parallel dichotomy in the study of psychosis, of art, and of society. This is the dichotomy of objective and subjective, cause and meaning, universality and relativism. Psychosis may be reducible to a few biological mechanisms and diagnostic categories, but the actual experience, the subjective *content* of delusions and moods, is uniquely imaginative and entirely personal. Art, as a behaviour is, if Miller is right, universally a matter of sexual display. No great difference between individuals or between societies there. However, the meaningful *content* of art is always different, always changing, always exceeding all previous meanings in some way. Similarly, much of the everyday behaviour of human beings in their societies, at a gross level, may be reducible to some universals of motivation, but the *meanings* that are attached to that behaviour are vastly different across time and space, which is what makes humanity so fascinating.

Thus both the scientific, universalist position, and the humanistic, relativist position are correct in their ways. We humans are all basically the same, but at the same time we are all always coming up with different stories about who we are. The reason we do this is that we have been sexually selected to be *Homo imaginans*, the entertainer of fictions and fables, the spinner of yarns, and once we have spun those yarns we tell them to each other, always varying and twisting them and introducing something new. Thus for humans as no other animals, there is a gulf between objective and subjective, action and thought, cause and rationalization, behaviour and meaning. Behaviour has a kind of terrible practicality to it, whereas the stories we weave about what we are doing are literally fabulous. There is a link, of course, between what we do and the stories we tell ourselves about why we do it, but it is not a completely simple one. Our imagination interposes itself, and a great source of discomfort in mental life, as in culture, is discovering the gap between what you are doing and what you imagine you are doing, between who you are and who you imagine yourself to be. The gap between the mundane reality of what we actually do, and the imaginative story we tell ourselves about what we do, was well captured by William Faulkner in his novel *As I Lay Dying*, where he wrote about:

> ... how words go straight up in a thin line, quick and harmless, and how terribly doing goes along the earth, clinging to it, so that after a while, the two lines are too far apart for the same person to straddle one to the other.

People who go mad are people for whom the two lines become completely separated. The most creative individuals in our cultures are those who can straddle the most distance, and yet keep the two lines together. Such tricks hath strong imagination; such a creature is *Homo imaginans*.

CHAPTER 8

Civilization and its discontents

> No condition is permanent
> African proverb

In the course of this book, I have argued that madness is not a mere aberration caused by some nasty quirk of current socio-economic or environmental conditions. Instead, the capacity for madness is a fundamental part of human nature. It is inherent in our psychological make-up, and preserved in the gene pool of our species, and for good reason. It is preserved because the very traits that make madness possible also underlie one of the things that we as humans most value, namely enhanced creativity. If this thesis, which we can call the genetic-creativity argument, is correct, then we should expect psychoticism to occur with roughly the same distribution in all populations. This is because it is assumed to be very deep in the human genetic lineage, and the genetic variants underlying it are assumed to have reached their equilibrium frequency. We might thus predict that the rates of psychosis will be about the same in all places and at all times. This chapter examines whether this is the case, and if not, why not, and what it means for the genetic-creativity argument.

◆

In 1871, the eminent British psychiatrist Henry Maudsley read a paper to the Medico-Psychological Association entitled 'Is insanity on the increase?'. He concluded that it was, and he was not alone in expressing alarm at the trend. Admissions to asylums in England trebled between 1869 and 1900. They continued to rise sharply until the dismantling of the great institutions which began after the Second World War. Even allowing for better diagnosis and the general expansion of professional medicine, the problem of insanity appeared to be growing. What Maudsley was observing was a phenomenon that emerges fairly consistently from the admittedly scant scientific literature on the prevalence of psychosis: that is, the proportion of people recognized as psychotic at any one time tends to be higher in industrialized, Western cultures than elsewhere. This is just part of a wider picture of rising psychiatric unease: the incidences of minor depression, eating disorders, and other personality problems have also increased steadily in the West, generation on generation, particularly in the past fifty years. Thus an apparently similar pattern emerges from studying historical trends in psychosis and from studying similar trends in minor disorders. A similar pattern also emerges from comparisons across cultures.

Societies with a more traditional, less urbanized, way of life appear to have lower prevalences of psychosis. Rates are reported to be low in interior New Guinea, on Tonga, on Sumatra, and amongst Taiwanese aborigines. In the USA, rates are lower in rural than urbanized settings, and are lower in the traditional subculture of the Old Order Amish than the rest of the population. The modern way of life would seem to be implicated directly; in New Guinea and the Solomon islands, rates of schizophrenia are reported to be several times higher in coastal populations, which have extensive contact with modern capitalism, than in the interiors where life is more traditional. Amongst the Tallensi of Ghana, a traditional people beginning to be involved in the modern economy, the anthropologist Meyer Fortes reported that thirty years of increasing experience of such things as wage-labour, consumer goods, and economic migration had led to a huge increase in the incidence of psychosis. Findings such as these have led many authors, over the decades, to conclude that madness is specifically a disease of modernity.

This idea has a long lineage. For the French eighteenth-century philosopher Diderot, civilization had been bought only at the cost of our natural happiness, an idea taken up in Rousseau's picture of the noble savage. Later, Freud developed the view that civilized life required us to frustrate more and more of our basic urges, leading to an increase in neurosis, if not in psychosis. His tract *Civilization and its Discontents* remains one of the most influential indictments of the mental unhealthiness of modern life ever written.

The idea that civilization breeds mental illness has resurfaced in more scientific form in the recent work of evolutionary psychiatrists Anthony Stevens and John Price, who advance a view known as the 'genome lag' hypothesis. According to this hypothesis, our minds are adapted to a particular type of environment, namely that of small hunter–gatherer bands living on a savanna somewhere. This is known in the jargon as our *environment of evolutionary adaptedness*, or EEA for short. Our archetypal needs—for status and affiliation—and fundamental psychological capacities—to deal with loss, danger, stress, and uncertainty—were honed by evolution in the EEA. Unfortunately nowadays, the situation has changed. For several thousand years, most humans have lived in larger, settled agricultural communities. More crucially, for a hundred years or so, most have lived in vast, anonymous, industrial towns, where the human challenges and opportunities are completely different from those we were designed to cope with. Not nearly enough time has elapsed for evolution to adjust us to the industrial environment, or even the agricultural one for that matter, since a few thousand years is hardly a blink of the evolutionary eye. Our genome has not caught up with our circumstances. It is still making people for the EEA.

If there is an increase in madness in modern life, then, it is because we are ill-equipped for the situation we find ourselves in, and the situation we find ourselves in is ill-designed for fulfilling the needs we bring to the world. Thus we break down much more often. As Stevens and Price put it:

> It seems likely that the various neuroses, psychopathies, drug dependencies, the occurrence of child and spouse abuse, to say nothing of the ever rising crime statistics, are not unconnected with Western society's inability to satisfy the archetypal needs of our kind.

There are some obvious problems with the genome lag hypothesis. For one thing, Stevens and Price seem to imply that people living in their 'natural' environment, the EEA, would be happy, and so avoid such tiresome ennuis as psychopathology, crime, violence, and so on. To an anthropologist, this assumption is a cardinal sin. Evolution never said anything about making life harmonious or people happy. All evolution does is allow those best at reproductive competition to stay around while their competitors die out. Indeed, the central plank of evolutionary theory is that resources in the world are too limited for everyone to live peaceably. Life is a constant struggle, between predators and prey, hosts and parasites, males and males, males and females, even parents and children, because at bottom, our genetic interests all differ and the world ain't big enough for all of us. Evolution blindly tries out anything that might just work out in this dog-eat-dog world, and very often, things do not. Thus we should expect, even in the EEA, a loser for every winner, a behavioural strategy that is disastrous for every one that works, and a fair dose of violence, antisociality, and alienation to boot. Most people's archetypal needs were probably never met. Indeed, if they were, they had not been set high enough, since those needs are there to make us strive to win out in reproductive competition. There has never been a guarantee from nature that they would be met. Homicide rates in traditional societies were many times higher than in even the most violent cities today, and warfare, though less lethal, was more frequent. The assumption that widespread unhappiness and conflict must be a modern aberration just does not wash. Unhappiness and conflict were there in the game plan from the beginning.

There is a related problem for the genome lag hypothesis, which is that, in some respects, the environment has not really changed that much. Although we live in anonymous industrial towns, the number of people we actually interact with on a daily basis is still fairly small. Departments in large firms, circles of friends, and social clubs are still, by and large, of a similar size to the bands of hunter–gatherers, and it is probably no accident that this is so. More importantly for present purposes, the major stressors that precipitate psychiatric illness are remarkably enduring: death of spouse, romantic rejection, loss of relatives, childbirth or loss of children, loss in professional or political competition. These are the fundamental problems of status and attachment that always have, and always will, follow us from the cradle to the grave. Exactly the same issues have been preoccupying people since time immemorial, and only the last one—professional competition—is very much different now from how it was back in the EEA. Nowadays we compete for status on the trading floor, on the pay scale, in college examinations. These situations are all novel, yet the fundamental dynamic is still much the same. Substitute clinching the big deal for bringing home the big bison and you will find that not much has changed.

Despite these problems, the genome lag hypothesis has a fundamental attraction to it, which is that it accounts compellingly for contemporary increases in the prevalence of psychiatric disorder. I believe it can be stated in a way that makes evolutionary sense, and I will attempt to do this. However, there is a more difficult issue to be considered, which is that it seems to conflict with the main thesis of this book. The genome lag hypothesis is a 'bad environment' hypothesis. A bad environment—modernity—causes a basically well-adapted organism to break down. By contrast, I have argued that the fundamental mechanism behind psychosis is 'bad genes', genes that have been preserved in the gene pool because of their compensatory advantages. Psychoticism is therefore built into the system of humanity, and would be expected to appear *whatever* the environment. It was there in the EEA, because you can't have the benefits of enhanced creativity without it.

Once again we seem to have a stand-off between a nurture position and a nature position, bad environments and bad genes. I will argue that no stand-off is really necessary, though both positions contain some truth. On the one hand, the genetic case is watertight, and the distribution of psychosis is consistent with the idea that it is a universal part of the profile of humanity. On the other hand, rates of mental illness change significantly over time and across cultures, so environments do make a difference. There is a path through this minefield which respects the truths in both positions, and it is this path that I will try to trace in this chapter.

◆

Let me deal first with the evolutionary problems of the genome lag hypothesis. Now it is true that we should not expect people in their EEA to live happy, stress-free, harmonious lives. We should expect them to be just as stressed and prone to unhappiness and disappointment as anyone else. However, we should expect to them to respond to stresses and disappointments in a generally adaptive way. That is to say, faced with broken attachments or failure in status competition, we might expect them to pick themselves up and make the best of a suboptimal situation. This is what those with psychiatric problems singularly fail to do. Faced with professional failure, they depress and thus fail even more. Faced with unfavourable beauty comparisons, they binge or starve themselves, and become even more unattractive. What is more, they endanger their own basic life processes. This cannot be part of the game plan. In short, in the EEA, we might expect violence in the service of biological goals, we might expect unhappiness, which is there for good reason, since it directs people away from things that are bad for them, and we should certainly expect stress, which readies people for reproductive competition. But we should not expect self-directed or futile violence, depression, or stress-induced breakdown. People would have been selected to be resistant to these.

So can we identify features of modern environments that might especially try our capacities for dealing with unhappiness and stress? The British psychologist Oliver James has argued cogently, in his book *Britain on the Couch*, that we can. Problems of low mood, he reminds us, can arise when we compare ourselves to others with respect to status or affiliation. Recall that subordinate vervet monkeys have lower serotonin levels than dominant ones, who have higher status and better relationships with the females of the group. In traditional societies, we would have been able to compare ourselves to perhaps a hundred and fifty other people, and know our status in their world. This no doubt led to many people having low mood, and low serotonin levels, but perhaps not pathologically so. However, nowadays, thanks to the mass media, we have thousands if not millions of people we can compare ourselves to. We are bombarded with images of abnormally beautiful, implausibly successful, creative, rich, attractive people, through advertising, particularly on television. Naturally, we make tacit comparisons with our own more prosaic existence. Modern capitalism preys on our desire to keep up with the Joneses by inventing ever more beautiful goods to have, places to visit, sports and cultural forms to master, and occupations to excel in.

For women, one of the traditional domains of status competition has been physical attractiveness. For them, the barrage of images of beautiful people not only fuels a multi-million dollar cosmetic industry, but must be related to the rising tide of eating disorders and associated depression. And this is not helped by the fact that women are now simultaneously, and quite rightly, encouraged to compete in other domains, too. As for men, traditional domains of male–male competition have been translated in our society into the quests for money, power, and cultural eminence, quests which are getting ever more difficult. Instead of competing with fifty other men, we are competing with thousands, if not millions, and also competing with women, who are often better at it than we are.

As for attachment, it is true that the fundamental issues—finding a mate, relations with kin and children, and the esteem of peers—are still the same as in the EEA, but the great increase in geographic mobility makes this more difficult and unstable than ever. The breakdown of the extended family, precipitated by changing residential patterns, increased wealth, and greater individualism, is surely one of the most significant changes in this regard. We now have fewer close attachments (including fewer children) than ever, and thus expect proportionately more from each one. This may be too much to expect, if the ever-increasing divorce rate is anything to go by.

Thus I can restate the genome lag hypothesis as follows: we are equipped to deal with stress, competition, and difficulties of attachment, since these fundamental problems of living were there in the EEA, but modern society may *overload* our equipment for coping positively, since it intensifies the problems to an unusual extent. More and more of us thus fall into patterns of self-destructive behaviour which lead us into the arena of medical concern.

This account makes sense, and certainly explains, for example, the great increases in depression and related disorders seen in the West since 1950, but how can it be squared with the rest of this book? Well, the big increases in mental disorder since 1950 are mainly seen in the minor disorders. Minor depression and the related serotonergic problems of antisocial behaviour, obsessive-compulsive disorder, anorexia, bulimia, addictions, and compulsions, have all seen big rises. This is understandable, since these can all be seen as problems of low serotonin: low serotonin arises from feeling you are low down on some hierarchy, and intensified social comparison makes more people feel this way. The psychoses, on the other hand, show far less tendency to increase. There is no increase in bipolar affective psychosis, and an increase only appears in unipolar illness when the category is drawn fairly broadly. Similarly, the lifetime risk of schizophrenia has shown no increase since 1950.

One way of squaring the genome lag hypothesis and the genetic argument, then, is to distinguish minor from major disorders. The discontents of civilization are precisely what Freud said they were: the neuroses, or minor disorders. These are the problems treated with psychoanalysis, Prozac, and better social conditions, which are obviously tied to life situations, and which have generally increased in recent decades. Those disorders fundamental to the human condition, which run in the genes and have nothing to do with modernity, are those treated with lithium, valproic acid, and antipsychotics, and which seem to have a more implacable aim.

I should beware of making this distinction seem too watertight. Much of the early part of this book was devoted to demonstrating that in both mental illness and the personality traits that underlie it, there are only continua, not natural breaks. Minor disorders are often found in the families of those with major disorders, and may thus be the biological penumbras of the major psychoses. Their occurrence indicates moderately high levels of psychoticism. The increased stresses of modern life are probably dragging people with lower and lower vulnerabilities into the arena of minor mental illness. People whose levels of vulnerability would have been low enough, in a previous era, to go through life quite normally, are now suffering depression, especially if they end up in an unfavourable point on the social heap.

If this is so, then why have the incidences of the major psychoses not increased at the same rate as those of the minor disorders? The simplest explanation is that for a major psychotic episode, you need to be *very* high on the relevant personality dimension. It may be that you need a difficult environment too, but without the strong predisposition, no environment will elicit the extreme psychotic outcome. Rates cannot therefore increase that much; they cannot outstrip the population frequency of the genotypes, and, normal distributions having the bell shape that they do, the very extreme genotypes are pretty rare.

♦

Now the reader may feel some confusion at this point. I began the chapter with the observation that insanity increased with modern socio-economic development, and I introduced the genomic lag hypothesis as a way of explaining this. I then argued that the genomic lag hypothesis accounts for the increased prevalence of minor psychiatric conditions in modern societies, but does *not* apply so strongly to insanity, which is universal and genetic. This is all very well, but Maudsley's observation that *insanity* was on the increase still needs explaining. Now I will show that, for psychosis itself, the universal, genetic argument, and observations of variability in time and space, can both be right in a way.

Good studies of the rate of psychosis across time and space, using standardized diagnosis and sampling techniques, are sadly rare. For schizophrenia, there has been a series of careful studies carried out under the auspices of the World Health Organization in 10 countries in the West, the Third World, and the former Soviet Union. There are also around 70 other studies, of variable reliability. For affective psychosis, there are a large number of epidemiological studies, though they are mainly from the West, China, and Taiwan, and they use highly variable criteria for diagnosing the disorder. Although these studies are the best of their kind, the data in them must still be considered warily, since cultural taboos about psychiatric conditions, as well as norms of health and disease, vary widely, and this can lead to misleading comparisons. That said, I will, for present purposes, take the studies at face value.

The World Health Organization studies of schizophrenia, and those studies of affective disorder that narrow attention to the most serious forms, especially bipolar, score an early point for the universal, genetic case. Psychosis is known in all societies that have ever been studied systematically. This itself is a very important finding, and supports the genetic-creativity argument. Furthermore, there is nothing in the data that forces us to reject the idea that the psychotic genotypes occur at approximately constant frequencies in all populations. The incidence of schizophrenia in the WHO studies is around 1–4 cases per 10 000 people per annum; this amounts to a lifetime risk in the order of 1 per cent, and the margin of difference between the 10 countries is not sufficient to suspect that much other than sampling error and diagnostic uncertainty is at work. Similarly, for bipolar affective disorder, the lifetime risk is a little under 1 per cent, with remarkable cross-national consistency. For the unipolar form of the disorder, figures are a little more variable, with a mean about 4 per cent. The variability here is generally to do with how much minor depression is included along with the major cases.

Round one to the universal, genetic case, then. However, this is an argument of two halves. As I have noted, the prevalence of insanity *did* increase in the nineteenth century, is higher in developed and urbanized societies, and remains low in groups like the Amish who retain a very traditional and stable social structure. How can these two findings be accommodated?

For one thing, the rate at which the genotype progresses to the disease is a variable. Recall that twin and family studies in the West show that only about 50 per cent of people with a strong genetic predisposition to psychosis actually develop it. Let us call this 50 per cent the conversion rate. We know that conversion is affected by social variables such as conflict, loss, failure, and so on. If one social set-up was more 'toxic' in terms of these stressors than another, then its conversion rate might be considerably higher. On the other hand, there may be societies that provide social roles and social support for vulnerable people so effectively that their conversion rates are extremely low. This could produce differences in the rate of psychosis of several-fold, without the population gene frequencies being any different. The differences between highland New Guineans, with their traditional social structure, and coastal New Guineans, coming to terms with modern capitalism, or between the Ghanaian Tallensi before and after they encountered wage labour, may well be explained in these terms. An increasingly difficult environment of social competition and disruption increases the conversion rate; some societies are better at making people ill than others.

Conversion rate differences are not the only societal factors affecting the rate of psychosis. Studies that look at the *proportion* of the population that is psychotic at any one time find a much greater degree of variation across societies than those, like the WHO studies, that look at *incidence*. In a recent review of 70 studies, E. Fuller Torrey found a range of prevalences of schizophrenia from 1 person per thousand to around 10 per thousand. These differences are affected by differing life expectancies; in a population where life expectancy is low, there is a relatively high proportion of children, and since schizophrenia is a disease of adulthood, this artificially lowers the proportion of schizophrenics. Correcting for life expectancy effects, however, produces an even larger range of variation, from around 1 per 1000 in Ghana, to 15 or 20 per 1000 in parts of Sweden.

Why should prevalence be so much more variable than incidence? The only conceivable explanation is that there are differences in the course of the disease, and the WHO studies confirm that this is indeed the case. In the developing countries studied by the WHO investigators, the disease profile of the schizophrenics differed from that found in the West. Third World patients were more likely to have a sudden onset, a brief duration of psychosis, and, crucially, were more likely to recover completely than their Western counterparts. Two years after initial diagnosis, only 24 per cent of British and 23 per cent of US schizophrenic patients had a 'very favourable' outcome. This means that they had returned more or less to their pre-disease lifestyle. By contrast, 57 per cent of Nigerian patients and 49 per cent of Indians fell into this category. This confirms the picture from other research. One study in Mauritius found that 59 per cent of schizophrenics had recovered completely 12 years after first hospitalization; the comparison figure for Great Britain was only 34 per cent. In all of these studies, the percentage of people who have a very unfavourable, chronic outcome, and require permanent medical care, is relatively constant. The difference is that more people in the developing countries have a quick and complete recovery, as opposed to some intermediate, lingering, partially recovered outcome.

This effect is sufficient to account for the differences in prevalence across societies. If the rates of new cases in societies A and B are the same, but in society B people get better more often and more quickly, then the proportion of people who are psychotic in any time period will be lower in B. Society B can have a much lower prevalence, even if the gene frequencies and the conversion rate are exactly the same. Thus there is another dimension which is variable in time and space: some societies appear to be better at making people better than others. This may well account for the apparent increases in insanity in nineteenth-century Europe. Rapid urbanization, the disruption of traditional agricultural communities, and the breakdown of the extended family, left generations who didn't get better, had nowhere to go, and who thus swelled the wards of the new asylums.

It is of the greatest interest to consider what it is about less developed societies which makes them better at making people better. Certain factors are easy to identify. For one thing, there are strong traditions of extended family cohesion. Such networks of relatives may be better at accommodating a person who is not entirely well than our more individualistic, nuclear family situation. Thus a person who is no longer frankly psychotic but still poorly adjusted is more likely to be taken back into the fold, and thus falls out of the view of medical statistics.

(This doesn't mean that he is really better, of course, merely that society has a capacity to absorb him.) A reliably supportive home situation may well help the rehabilitation process, though, and the spreading of the burden over a large number of relatives (more children, as well as more siblings, cousins, uncles, and even wives) makes the emotional environment much less intense and stressful. Secondly, in traditional societies, there are ready-made economic roles which are psychologically undemanding and for which the convalescent does not have to compete. Going back to work on the family farm, or the family market stall, is less difficult than going back out into the job market and justifying the gap on one's CV to a complete stranger. The social competition that the recovering psychotic must face is once again much greater in modern capitalist societies. These factors are worthy of a great deal of further study, since they might point to the things that would best support recovering psychotics in the West. Underdeveloped societies have lots of undesirable features, but if they are good at psychological healing, then this is an area where we could try to learn from them.

◆

The conclusion of this chapter, then, should be a positive one. On the one hand, psychosis is universal, as the genetic-creativity argument requires. On the other hand, we do not seem to be completely stuck with it. As a society, we can make a difference, both by minimizing the rate at which the genotype gets converted to the disease, and by fostering the conditions that make people better. These are important and constructive lessons, and they carry forward to our consideration, in the final chapter, of what can and should be done about psychosis.

CHAPTER 9

Staying sane

> It isn't danger; it's not an accomplishment. I don't think it is a visitation of the angels, but a weakening of the blood.
>
> Robert Lowell

Let me restate the genetic-creativity argument one final time. The major psychoses are basically genetic in origin, though the genetic liability may need triggering by unfavourable aspects of the environment. The genes that underlie psychosis affect the personality trait of psychoticism, by coding for aspects of brain chemistry and anatomy. These personality traits are not wholly negative. As well as madness, they are associated with great creativity. Great creativity is highly attractive, and so the traits, and the genetic variants behind them, persist in the gene pool.

Madness and heightened creativity are thus the very different edges of the same sword. This has been shown by several lines of evidence: the rates of psychosis in creative professionals and their families; the similarities in cognitive style between psychotics and highly creative individuals; and finally, going way back to the Introduction, the similarities in form of psychotic delusions and literary creations.

The genetic–creativity argument has, amazingly enough, two parts: the genetic part, and the creativity part. The reader may be tempted to draw a sweeping conclusion from each part. From the genetic part, he may conclude that psychosis is inevitable. After all, if something is determined in the genes, then surely nothing can be done about it. From the creativity part, he may conclude that psychosis is not an unambiguously bad thing. It increases creativity, which is a good thing, so perhaps we shouldn't be trying to do anything about it. Surely then, we have to grip the double-edged sword firmly in our hands, both as a society and as individuals, first because we can't avoid it, and second, because we shouldn't want to avoid it.

This final chapter is devoted to showing why these conclusions are deeply wrong. Firstly, we know from the evidence of the previous chapter that societies are not stuck with their levels of psychosis, despite the undeniable involvement of genetic factors. Secondly, I argue that, despite the creativity association, we should want to fight psychosis, and that as a society we can do so, at several levels. The fact that nature plays a role does not mean that nurture is not important. It actually makes getting nurture right much more important. At the individual level, this means that people with a vulnerability to psychosis should not embrace their predestined trip; rather they should grab hold of every tool they can to protect themselves, including, most importantly, living a healthy life, just as fiercely as those burdened with a predisposition to cancer. Thirdly, I argue that those with creative ambitions should not seek out psychotic experience; this is a foolish, romantic notion, and it will do them no good at all. Whatever you want to achieve, staying sane is the best way to achieve it.

♦

Is psychosis inevitable? We saw in the previous chapter that it is not. Societies can get better at making it better, and if some combination of developing countries' supportive social situations and modern psychiatric treatment could be found, we could get a lot better at making it better. But we must also consider whether it is inevitable that psychotic breaks should happen in the first place. Though there is an inherited genetic liability, the rate of conversion of the liability into the disease stands, at present, at around 50 per cent. In earlier chapters, I advanced this 50 per cent figure as strong support for the involvement of genetic factors in psychosis, and indeed it is. However, you can also look at it another way. Even with the full genetic predisposition, your chances of becoming psychotic are only one in two, which makes managing the factors that determine which half you end up in very important indeed. The strong role of nature makes managing nurture *more* important, not less, because it can make such a striking difference.

As yet, we do not entirely understand what the factors are that determine which 50 per cent people fall into, but they certainly include triggering life events, and childhood sensitization. According to the threshold model, people with a very strong genetic predisposition do not require much by the way of triggering or sensitization to suffer a psychotic break, but none the less, they do require some environmental input, and we simply do not know how fixed the 50 per cent conversion rate is. There is no theoretical reason to believe that it is a constant, so perhaps it could be brought down, both at a societal and an individual level.

Drugs have an important role to play in this, particularly in affective psychosis. The major medications for affective disorder are generally fairly effective, both in treatment and in prevention. Lithium and valproic acid (used for bipolar patients) and antidepressants (used for unipolar patients) all have an important maintenance function. That is to say, they are prescribed not just when the patient is ill, but also while he is healthy, to keep him healthy. It is easy to see how this works, as these drugs provide a constant chemical support to the patient's unstable serotonergic metabolism. In general, these drugs are not prescribed until the first psychiatric episode, for two reasons. The first is that vulnerability does not become apparent until the first episode, and the second is that all of the drugs are fairly 'dirty', which is to say that they have a wide range of chemical effects beyond the desired ones. This is not surprising, considering that they were all discovered by accident. New generations of rationally designed psychiatric drugs, to which Prozac pointed the way, will be cleaner and more specific in their actions, lower on side-effects, and less toxic in general. This, coupled with greater public awareness of risk, and perhaps with genetic screening, should mean that many people with a liability to affective disorder should be able to go through life without any episodes at all.

For schizophrenia, sadly, the picture is not quite so rosy. The positive symptoms of schizophrenia often respond to treatment with antipsychotic drugs. However, the troublesome negative symptoms and cognitive impairments are more intractable, and it may be these that cause long-term problems of social and economic integration. The proportion of schizophrenics making a full recovery in developed countries has not changed significantly in a hundred years, despite the development of drugs. A drug that could do for schizophrenics what Prozac has done for minor depressives and lithium has done for bipolar patients is surely a Holy Grail for psychopharmacologists. However, though our understanding of the brain mechanisms of the disease advances apace, the diffuse and intrinsic nature of those mechanisms makes the discovery of a magic bullet seem unlikely.

Although I mention the importance of the drugs provided by the big drug companies, I should not neglect the even more crucial drugs that we ourselves control. We have the power to self-administer powerful serotonin-enhancing chemicals every day; we do that by living a good life, which suits us, and which we enjoy, and by seeking out rewarding situations and friends. We cannot entirely control the stressful things that may befall us, but we can develop resources for coping with whatever life throws at us, and develop positive frames of mind and support systems which will make them will seem more bearable. The talking cures of therapists and counsellors are drugs of this kind, and for minor conditions they can perform comparably with the drugs that come in bottles. They may be even more helpful in making us stronger in the long run. For major psychoses, talking cures are most useful as tools for rehabilitation once conditions have been stabilized.

Individuals at high familial risk for psychosis can thus help themselves. They, like all of us, should manipulate their own environments to maximize psychological health. That is, they should choose their surroundings, their companions, and their professional challenges very carefully, if need be with professional help from counsellors and psychotherapists. The golden rules are: be honest, realistic, and loving with yourself and those around you; assume a positive outlook and make brave, positive life choices, going against the grain where necessary; listen to your body and keep healthy and active; and do not tolerate niggling discontent, whatever its cause, even for a short time, and even if it seems minor. Actually these are rules that everyone should follow at all times, but for the member of a psychotic family they become particularly important. Being unhappy is a strong predictor of later mental and physical illness. The niggling discontent will always out; but whereas it might lead to a broken relationship or an unhappy career for most people, for the vulnerable person, it can destroy him and those who love him in the most terrifying way imaginable.

This is not just an issue for individuals. Communities and social networks which are fragmented, stressed, or in which there is no sense of belonging have been shown to lead to poorer outcomes and higher suicide rates for the mentally ill. This is something that social policy can work on. The existence of a biological predisposition thus in no way relieves us of the responsibility for doing something about mental illness.

Am I not getting ahead of myself here? I have asserted that positive steps can be taken to evade psychosis, assuming that such an evasion is the right goal to have. But, looking back over this book, this may not be self-evident. Have we not seen that psychosis is linked to great creativity? However unpleasant, perhaps it is a small price to pay for a heightened life of the mind. Have we not seen several distinguished psychotic authors opine that in the storm of their madness there was a glimmer of special vitality, courage, and insight? Perhaps that glimmer compensates for the terrible costs of their condition. Surely the society that tried to eliminate psychosis by drugs or any other means would be a society of dull conformity and deadened vision, which deprived itself in the long term of the best and most vital in human culture, which valued the ultilitarian drudge over the possibilities of the imagination. Similarly, the individual who chose a drugged calm over a difficult but creative tempest would surely be a coward, who had failed to live true to himself and be the most excellent human he could have been.

We might call this the romantic position. It is romantic in a double sense; at one level, it is obviously a romanticization of psychosis, and a refusal to condemn it. More specifically, it draws on the notion central to European culture since the Romantic movement that extreme and tempestuous experience has a special value—virtue even—that more banal activities lack. The romantic position is most easily endorsed by someone who has never been close to a psychotic. It is emotionally appealing, perhaps; heroic, certainly; and wrong, utterly.

The first point to make is that the romantic position understates the negative aspects of psychosis, and overstates the positive. We should never forget for a moment that the psychoses are severe, crippling, often lethal diseases. They destroy careers and marriages, and frequently lead to suicide. Moreover, in most cases of psychosis, there is no silver lining in the black cloud, either for the person, their loved ones, or their culture. Most psychotics never achieve anything. They are destructive of everything they have built for themselves. They have the same old trite, hackneyed, self-absorbed delusions; their behaviour is disorganized and injurious, and not directed into great creativity, nor indeed anything worthwhile at all. There are, it is true, a few remarkable individuals who achieve great insight in the midst of a psychotic tempest. However, these are exceptional individuals. They are unusually intelligent, disciplined, perceptive, positive people, who would probably have achieved great insight into their experiences *whatever* those experiences had been. Thus what sets them apart is not psychosis per se.

More generally, the romantic position confuses *psychosis* with *psychoticism*. Recall that psychosis is actual madness; psychoticism is the personality dimension which predicts, among other things, the predisposition to psychosis. People who are high on psychoticism do indeed have enhanced creativity. However, the creative work done by people who also suffer psychosis is not done during the psychosis itself. Psychotic episodes are almost always entirely unproductive, even in the most creative individuals such as Schumann. This is because the mind becomes too disorganized to plan and execute anything, least of all a creative work. Schumann's great output was achieved in the milder, hyperthymic periods *between* the terrible depressions, not in the storm itself. This is the normal pattern for those suffering affective disorders. By the same token, the people who have the highest creativity scores of all in psychological studies are not psychotics but the non-psychotic relatives of psychotics. That is to say, those who bear some of the traits of psychoticism but can avoid the actual psychotic breakdown do better than those who suffer the storm.

This partly explains the paradoxical personality profile of the most creative people. They tend to score highly on measures of psychoticism. However, they also score very highly on scales of intelligence, self-discipline, industriousness, seriousness, organization, and time management. Moreover, they are high on 'ego strength', a measure of internal control which predicts strong *resistance* to mental breakdown. These are not the traits usually associated with madness. Nor are they romantic (or Romantic) traits, but it seems that they are key to actually getting things done. The optimal combination is one of fire in the mind, along with other resources which can control and harness that fire. Fire is nothing without control.

The implications of these facts for creative output are clear. While psychoticism may be desirable in our culture, and enviable in ourselves, psychosis itself is not. Creative achievement will follow from harnessing the positive features of the underlying genotype and, if possible, eliminating the actual madness altogether. Thus professional writers should have no compunction about taking their medication. The few studies that exist suggest that psychiatric medications can actually increase artistic productivity, rather than decreasing it, by stabilizing the extremes. Admittedly, lithium, long the drug of choice in bipolar disorder, has some negative side-effects. The mood may be so thoroughly stabilized that the person becomes apathetic. Lithium also tends to be sedating, and may have some adverse cognitive effects. However, these are issues for more sensitive dosing, and eventually for more rational drug design. They are not an argument for trying to live unmedicated. In one study, most writers and artists stated that their productivity had either increased (57 per cent) or stayed the same (20 per cent) while on lithium, and lithium's successors will be better still. Compare this to the alternative; as Kay Jamison has pithily put it, 'No one is creative when paralytically depressed, psychotic, institutionalized, in restraints, or dead because of suicide'. The double-edged sword, then, is not psychosis. There is only one edge to psychosis, and it is bad. The double-edged sword is psychoticism, and that is quite a different matter.

◆

These considerations also have implications for those who aspire to creative pursuits. One might think, if psychosis is associated with creativity, then those interested in creativity should seek experiences of a psychotic nature, perhaps by living turbulent lives, forcing themselves into extreme emotional turmoil, or using dopamine-related drugs, which can mimic psychosis, such as amphetamines or cocaine.

Once again this is a romantic (and Romantic) error, and

rests on a confusion between psychosis and psychoticism. Heightened creativity comes from psychoticism; psychosis does too, but it does not follow that by mimicking psychosis you tap any of the other benefits of psychoticism. An analogy is useful here. There is a human genetic variant, the sickle-cell gene, which is common amongst Africans, and gives the bearer enhanced resistance to malaria. However, because the gene operates on red blood cells, when present in double dose it increases the risk of anaemia. This gene is thus the common fount of malaria resistance and anaemia. Now what would we say if a person lacking the gene tried to improve his malaria resistance by making himself anaemic? (He could do this by, for example, restricting his diet.) We would have to conclude that his strategy was completely absurd. He is just going to be ill and miserable and no nearer to having a sickle-cell genotype, or to avoiding malaria.

The creativity–psychosis case, though not identical, is in some ways similar. The genes for psychoticism have two effects: madness and enhanced creativity. Drugs like cocaine, and certain lifestyles, can mimic one of those effects (madness), but they do nothing for the other one (creativity), and of course they bring the user no nearer at all to being high in psychoticism. Nothing can do that.

Now it is true that a strikingly high proportion of artistic individuals have lived unorthodox lives, and used recreational drugs to regulate their output. In Arnold Ludwig's biographical study of eminent people, artistic types had three times the rate of alcoholism and drug abuse as other eminent people, a rate many times higher than that of the population at large. We know why people vulnerable to mental disorder might use drugs, since such substances change the balance of brain neurotransmitters. Addicts are thus effectively self-administering sedatives or dopamine rushes to counteract some of their intrinsic neurotransmitter imbalances. Affective patients who use drugs are self-medicating to try to quell the storm within, so that they can get on with their work. That is to say, drugs are used to ease the worst *dis*contents of the personality type, not because they aid creativity in any positive way. If you don't have that personality type, then these substances will not help you. Even if you do, they are not an ultimately effective therapy, and you should see a doctor for a better one. So if you aspire to creativity, the advice is clear. If you have a predisposition to psychosis, avoid it like, well, the plague. If you do not have such a predisposition, then do not pretend you do or go around wishing you did. This seems unhelpful advice; what can you do, more positively, to improve your creative performance? There are in fact positive lessons to be learned from the psychology of madness and creativity, and they are the following.

Firstly, for successful creative performance, psychoticism must be combined with other qualities (and it rarely is, which is what makes outstanding people outstanding). These other qualities are intelligence, obviously, but also hard work, organizational skills, commitment, and the willingness to keep trying. It might not seem very romantic (or Romantic), but most professional writers get up pretty early in the morning and get to bed on time, most musicians put in 6 hours practice a day, and most visual artists work long weeks. All these people have to drive themselves on through years of relative tedium, frustration, and low return, which requires self-discipline, humour, and resilience. If you want to create, you cannot change your personality type, but you can work on some of these other qualities. You can keep your mind in shape, develop firm goals and good routines, and above all, practise your craft, not just sporadically, but doggedly.

Secondly, we can identify what it actually is within psychoticism which is psychologically important for creativity, and if we haven't inherited the whole package (which we should be thankful we have not), we can at least work on approximating those things.

For the thymotypy subdimension, it is clear that the benefit of the genotype lies in the ability to generate a strong positive mood for the period required to produce creative work. As we have seen, affective patients are unrealistic in their moods—unrealistically pessimistic in depression, and unrealistically optimistic in mania. The high mood of mania is too high to be productive or in any way positive, but the good mood that lies a little way below it is associated with great energy to work, and optimism that tasks can be completed. This is immensely valuable for the long, frustrating period of development intrinsic to any worthwhile creation.

High mood is not the preserve of manic-depressives. They get it for free, as it were, since their unstable serotonergic system throws them into it easily but intermittently. However, the rest of us can get it as regular paying customers. The ways we buy into it have been trumpeted in a thousand manuals of self-help psychology, and as a result are almost clichéd by now. However, there is also a scientific psychology of happiness, which confirms empirically all the main exhortations of these manuals. We can improve our mood by knowing ourselves, allowing ourselves pleasure, making a good environment to live in, seeking out the things we really want to do, setting the goals towards them realistically, and pursuing them with determination. We can also help ourselves by staying fit, healthy, sociable, and positive. Such strategies are nearly free, though they require some hard work at the outset, and they are of almost limitless value, whatever it is we want to achieve. Experiments show that a good mood reliably makes people quicker, more accurate, and more effective on a wide range of cognitive tasks.

As for the schizotypy subdimension, the benefit lies in the ability for unorthodox, divergent, lateral thinking. This obviously comes more easily to some people than to others. However, we can all cultivate it, by subjecting ourselves to a wide variety of experiences, and training different facets of our minds, whether through lateral thinking courses, mathematics, music, or art classes. The mind is not completely a done deal; it is also what you make of it.

The attributes I have discussed all seem very far from madness, where I began, and indeed they are. Madness is the antithesis of good mood and lateral thinking. But, as we have seen, it is a state with paradoxical underpinnings. Those underpinnings are oxymoronic; the superior mental order which leads to mental chaos, and the overwhelming joy which leads to overwhelming despair. Within the capacity for madness dwell some positive qualities which help make people more creative. We should not go seeking the whole package, but we should look, learn, and pluck the strawberry that grows beneath the nettle.

Epilogue

I have now reached the end of my journey through the world of psychosis. It took me to some unexpected places; from Kraepelin to Shakespeare, Schumann, Inuit songs, peacock's tails, genes, and neurotransmitters. The very variety of this journey affirms the strangely central place of madness in human experience, lying as it does at the intersection of individual psychology, medicine, brain science, genetics, evolution, culture, and literature.

In writing the book, I have been often struck by the awesome power of the impersonal forces that determine who we are. Will I create a masterpiece? Will I go mad? Not, it might seem, choices that are mine to make. My serotonin and dopamine systems make me who I am, and these things are determined by genes, which I did not choose. However much we kid ourselves otherwise, we are basically bunches of chemical reactions, the course of which is predetermined to a significant extent by our DNA. Similarly, our experiences in love and work will depend heavily on the effects of such early events as the loss of a parent, events we did not choose or control to any significant extent at all. It is all too easy to become gloomy about the prospects for changing oneself or humanity. Where is free choice in all this?

Gloomy conclusions are not warranted, though. I take great heart from the 50 per cent rule. Only 50 per cent of people with psychotic genotypes become psychotic. Many of the other 50 per cent live very satisfying lives, the course of which is not predestined by that aspect of their heredity. A person exceeds the sum total of all the influences acting on him, be they genetic or cultural. That is what makes him a person. Fifty per cent of the time, a given person might die a madman, but the other fifty, he might live a poet. Taking that 50 per cent chance, controlling that flip of the coin, is what being human is all about. We may not, in the long term, be stronger than the forces of nature and nurture, but while we are here we can give them a run for their money. Evolution, paradoxically, has given us the cognitive equipment to look at what it is doing with us, and by the insight thus gained, perhaps make it back off for a bit. Psychiatric medications, which will get better and better, are just one domain where humanity has built a tool with which to reach into the machine evolution has made us, and retune it to our own purposes. Not all such tools are hi-technology. We are using such a tool every time we decide on a new career, a new creative endeavour, or a new lifestyle. We are writing our own scripts, changing the balance of our own neurotransmitters, controlling, however briefly and partially, the flip of the coin that decides who we are.

Thus the overall determination of the life of the human species by the massive, impersonal forces of evolution sits alongside an underdetermination of what happens at the level of the individual human being, and this underdetermination gives every person a vital and fiery hopefulness. The poet Ted Hughes captured this tension very well in the poem *Examination at the Womb-Door* from his *Crow* cycle.

In the poem, Crow, his gawky, awkward hero, is being born. As he enters the world, the figure of Death is standing sentinel at the womb-door. Death reminds Crow in no uncertain terms just who is in control.

> Who owns these scrawny little feet? Death.
> Who owns this bristly scorched-looking face? Death.
> Who owns these still-working lungs? Death.
> Who owns this utility coat of muscles? Death.
> Who owns these unspeakable guts? Death.
> Who owns these questionable brains? Death.
> All this messy blood? Death.
> These minimum-efficiency eyes? Death.

> This wicked little tongue? Death.
> This occasional wakefulness? Death.

In the present context, we can think of Death as those long-term forces of biology that determine our fate. In this sense, it is Death, sure, that determines our allotted time on the Earth, physical capabilities, basic needs and drives, intelligence, personality, propensity to become mad, and so on. Whatever we try to do, however strong our humanity, we will hit the limits of temperament imposed by this biological frame in the end. Who owns these questionable brains? Death.

However, we can still affect when and how we meet those limits, and what use we make of those capacities. For the short time we are on the Earth, we can toss our own coin, or try to. We are, to that limited but vital extent, authors of our own destinies, and we should never forget this for an instant. This is how the poem ends:

> Who is stronger than hope? Death.
> Who is stronger than the will? Death.
> Stronger than love? Death.
> Stronger than life? Death.
>
> But who is stronger than death?
>   Me, evidently.

Pass, Crow.

Further reading

In this book, I have investigated the question 'Why does madness exist?' and a host of other questions about the mind, the brain, human behaviour, and evolution which turned out to be related to the central one. The presentation has been as concise and non-technical as possible. I have also avoided footnotes and off-putting citations in the text, although select references to the research I have discussed are listed in the 'References' section that follows this one. I have chosen this format because my aim has been to convey the important concepts of several areas of ongoing scientific research, and suggest how these concepts might fit together. It is my hope that the reader who is sufficiently interested may be inspired to pursue the technical details for himself. This 'Further reading' section gives some pointers for anyone wishing to start doing so.

The book covers six overlapping areas. These are as follows: the nature of mental illness in general; the nature of schizophrenia and affective disorder in particular; the brain mechanisms underlying mental disorder; the evidence of a genetic basis; the relationship of mental disorder to creativity; and finally, the relationship between mental health and society. I here list one or more key books for each of these areas. The books are generally written at a slightly more technical level than this one, although all are still accessible to the student or general reader.

On the nature of mental illness in general, Anthony Clare's *Psychiatry in Dissent: Controversial Issues in Thought and Practice* (2nd edition. London: Routledge, 1980) and Peter Sedgewick's *Psycho Politics* (London: Pluto Press, 1982) still make very interesting reading. They do appear a little dated, however, since they are largely concerned with responding to the 'anti-psychiatry' movement of the 1960s and 1970s, which attacked the whole 'disease and medicine' ethos of psychiatry. These debates have largely ceased, with the biomedical orientation triumphant, but such conceptual issues as the limits of the disease concept are still very pertinent. For an excellent scientific presentation of the dimensional approach to mental disorder, including the concept of the temperamental threshold, see Gordon Claridge's *Origins of Mental Illness: Temperament, Deviance and Disorder* (Cambridge, MA: Malor Books, 1995). On the history of psychiatry in general, encompassing all approaches, Michael Stone's comprehensive *Healing the Mind: A History of Psychiatry from Antiquity to the Present* (New York: Norton, 1997) cannot be bettered.

For the specific disorders, Irving Gottesman's *Schizophrenia Genesis: The Origins of Madness* (New York: W. H. Freeman, 1991) is the most informative single volume on schizophrenia. The state of the art of contemporary research on the neurobiology of the disease is reviewed in the articles in a special volume of *Brain Research Reviews* (volume 21, numbers 2–3, 2000). For the affective disorders, Peter Whybrow's *A Mood Apart: A Thinker's Guide to Emotion and its Disorders* (New York: Basic Books, 1997) is a thorough and readable introduction to all aspects of symptoms, research and treatment, and Paul Gilbert's *Depression: The Evolution of Powerlessness* (Hove: Erlbaum, 1992) is useful on depression more particularly.

On the brain mechanisms of psychosis, there is no volume with complete coverage. Samuel Barondes' *Molecules and Mental Illness* (New York: Scientific American Library, 1993) presents what is known about the operation of neurotransmitters and their drugs in a lively way. More technical in aim, but still very accessible, is Stephen Stahl's *Essential Psychopharmacology: Neuroscientific Basis and Practical Applications* (Cambridge: Cambridge University Press, 1996). Neither of these books covers the evidence of neuroanatomical irregularities in psychosis. For this, the reader is referred to

Nancy Andreasen's article 'Linking mind and brain in the study of mental illnesses: A project for a scientific psychopathology', *Science* 275 (1997), 1586–1593. As for the genetic basis of psychosis risk, Samuel Barondes' *Molecules and Mental Illness* is informative, as is his more recent *Mood Genes: Hunting for the Origins of Mania and Depression* (New York: W. H. Freeman, 1998). The evidence for genetic factors in schizophrenia is usefully synthesized in Irving Gottesman's *Schizophrenia Genesis*. The role of genes in personality and behaviour more generally is explored by Dean Hamer and Peter Copeland in *Living With Our Genes: Why They Matter More Than You Think* (New York: Doubleday, 1998). Genetic research advances at such a pace that books, including this one, simply can't keep up to date; Hamer and Copeland's account of the *D4DR* gene research, for example, was already outdated by mid 2000.

Kay Redfield Jamison's book *Touched with Fire: Manic-Depressive Illness and the Artistic Temperament* (New York: The Free Press, 1993) is a fascinating portrait of affective disorder in the lives of writers and artists, with a useful summary of other research on creativity and mental disorder. Gordon Claridge, Ruth Pryor, and Glen Watkins discuss the lives and work of 10 psychotic authors in their *Sounds from the Bell Jar* (Basingstoke: Macmillan, 1995). Arnold Ludwig's *The Price of Greatness: Resolving the Creativity and Madness Controversy* (New York: Guildford Press, 1995) reports his painstaking biographical studies of the psychological characteristics of eminent people. In addition, on the creativity issue, Louis Sass's *Madness and Modernity: Insanity in the Light of Modern Art, Literature and Thought* (Cambridge, MA: Harvard University Press, 1992) is an extensive and at times breathtaking exploration of schizoid thought and its relationship to modern culture. Geoffrey Miller's sexual selection theory of the origin of creative intelligence is set out in his book *The Mating Mind* (Oxford: Heinemann, 2000).

Several thought-provoking books deal with the relationship

of mental health to wider social issues. Oliver Jones' *Britain on the Couch: Treating a Low Serotonin Society* (London: Arrow, 1998) examines why developed capitalist economies should be so prone to breeding minor mental discontents. Meanwhile Peter Kramer's *Listening to Prozac* (New York: Viking Penguin, 1993) is a fascinating examination of the social and personal issues that surround the existence of such effective personality-changing drugs as Prozac.

References

Introduction

Haslam, J. (1810). *Illustrations of Madness: Exhibiting a Singular Case of Insanity, Embellished with a Curious Plate.* G. Hayden, London.

Jamison, K. R. (1993). *Touched with Fire: Manic-Depressive Illness and the Artistic Temperament.* The Free Press, New York.

Lombroso, C. (1889). *L'homme de génie.* Alcon, Paris.

Rush. B. (1812). *Medical Inquiries and Observations on the Diseases of the Mind.* Kimber and Richardson, Philadelphia.

Chapter 1: From difference to disease and back again

Al-Issa, I. (1995). The illusion of reality or the reality of illusion. Hallucinations and culture. *British Journal of Psychiatry*, **166**, 368–373.

Alloy, L. B. and Abramson, L. Y. (1979). Judgement of contingency in depressed and nondepressed students: Sadder but wiser? *Journal of Experimental Psychology: General*, **108**, 441–485.

Andreasen, N. C. (1997). Linking mind and brain in the study of mental illnesses: A project for a scientific psychopathology. *Science*, **275**, 1586–1593.

Bentall, R. P. (1990). The illusion of reality: A review and integration of psychological research on hallucinations. *Psychological Bulletin*, **107**, 82–95.

Boyle, M. (1990). *Schizophrenia: A Scientific Delusion?* Routledge, London.

Brockington, I. F. *et al.* (1979). The distinction between the affective psychoses and schizophrenia. *British Journal of Psychiatry*, **135**, 243–248.

Clare, A. (1980). *Psychiatry in Dissent: Controversial Issues in Thought and Practice*, 2nd edn. Routledge, London.

Claridge, G. (1995). *Origins of Mental Illness: Temperament, Deviance and Disorder.* Malor Books, Cambridge, MA.

Depue, R. A., Luciana, M., Arbisi, P., Collins, P., and Leon, A. (1994). Dopamine and the structure of personality: Relation of agonist-induced dopamine activity to positive emotionality. *Journal of Personality and Social Psychology*, **67**, 485–498.

Drevets, W. C. *et al.* (1998). Functional neuroimaging studies of depression. *Annual Review of Medicine*, **49**, 341–361.

Gottesman, I. I. (1991). *Schizophrenia Genesis: The Origins of Madness.* W. H. Freeman, New York.

Kennedy, M. F., Javanmard, M., and Vaccarino, F. J. (1997). A review of functional neuroimaging in mood disorders: Positron Emission Tomography and depression. *Canadian Journal of Psychiatry*, **42**, 467–475.

Knutson, B. *et al.* (1998). Selective alteration of personality and social behaviour by serotoninergic intervention. *American Journal of Psychiatry*, **15**, 373–379.

Posey, T. B. and Losch, M. E. (1983). Auditory hallucinations of hearing voices in 375 normal subjects. *Imagery, Cognition and Personality*, **2**, 99–113.

Sedgewick, P. (1982). *Psycho Politics.* Pluto Press, London.

Szasz, T. S. (1974). *The Myth of Mental Illness.* Harper and Row, New York.

Szasz, T. S. (1976). *Schizophrenia.* Oxford University Press, Oxford.

Wurtzel, E. (1994). *Prozac Nation.* Riverhead. New York.

Chapter 2: From nature to nurture and back again

Barondes, S. H. (1998). *Mood Genes: Hunting for the Origins of Mania and Depression.* W. H. Freeman, New York.

Bateson, G., Jackson, D., Haley, J., and Weakland, J. (1956). Towards a theory of schizophrenia. *Behavioral Scientist,* **1**, 251–264.

Freeman, D. (1983). *Margaret Mead and Samoa: The Making and Unmaking of an Anthropological Myth.* Harvard University Press, Cambridge, MA.

Gottesman, I. I. (1991). *Schizophrenia Genesis: The Origins of Madness.* W. H. Freeman, New York.

Laing. R. D. (1965). *The Divided Self: An Existential Study in Sanity and Madness.* Penguin, London.

Laing, R. D. and Esterson, A. (1964). *Sanity, Madness and the Family.* Penguin, London.

Mead, M. (1929). *Coming of Age in Samoa: a Psychological Study of Primitive Youth for Western Civilisation.* Cape, London.

Skinner, B. F. (1948). *Walden Two.* Macmillan, New York.

Chapter 3: This taint of blood

Barondes, S. H. (1993). *Molecules and Mental Illness.* Scientific American Library, New York.

Barondes, S. H. (1998). *Mood Genes: Hunting for the Origins of Mania and Depression.* W. H. Freeman, New York.

Benjamin, J. *et al.* (1996). Population and familial assocation between the D4 dopamine receptor gene and measures of novelty seeking. *Nature Genetics*, **12**, 81–84.

Bouchard, T. J. (1994). Genes, environment and personality. *Science*, **264**, 1700–1701.

Chapman, J. P. *et al.* (1994). Does the Eysenck Psychoticism scale predict psychosis? A ten year longitudinal analysis. *Personality and Individual Differences*, **17**, 369–375.

Chapman, L. J. *et al.* (1994). Putatively psychosis-prone subjects 10 years later. *Journal of Abnormal Psychology*, **103**, 171–183.

Claridge, G. (1995). *Origins of Mental Illness: Temperament, Deviance and Disorder*. Malor Books, Cambridge, MA.

Claridge, G. (ed.) (1997). *Schizotypy: Implications for Illness and Health*. Oxford University Press, Oxford.

Ebstein, R. *et al.* (1996). Dopamine D4 receptor Exon III polymorphism associated with the human personality trait of sensation seeking. *Nature Genetics*, **12**, 78–80.

Egeland, J. A. *et al.* (1987). Bipolar affective disorder linked to DNA markers on chromosome 11. *Nature*, **325**, 783–787.

Eysenck, H. J. and Eysenck, S. B. G. (1976). *Psychoticism as a Dimension of Personality*. Hodder and Stoughton, London.

Kapur, S. and Remington, G. (1996). Serotonin–dopamine interaction and its relevance to schizophrenia. *American Journal of Psychiatry*, **153**, 466–476.

Kempermann, G. and Gage, F. H. (1999). New nerve cells for the adult brain. *Scientific American*, May, 38–43.

Lesch, K.-P. *et al.* (1996). Association of anxiety-related traits with a polymorphism in the serotonin transporter gene regulatory region. *Science*, **274**, 1527–31.

Ogilvie, A. D. *et al.* (1996). Polymorphisms in serotonin transporter gene associated with susceptibility to major depression. *Lancet*, **347**, 731–733.

O'Rourke, D. H. *et al.* (1982). Refutation of the general single locus model in the aetiology of schizophrenia. *American Journal of Human Genetics*, **34**, 630–649.

Owen, M. J. (2000). Molecular genetic studies of schizophrenia. *Brain Research Reviews*, **31**, 179–186.

Paterson, A. D. *et al.* (1999). Dopamine D4 receptor gene: Novelty or nonsense? *Neuropsychopharmocology*, **21**, 3–16.

Rommens, J. M. *et al.* (1989). Identification of the cystic fibrosis gene. *Science*, **245**, 1059–1080.

Stahl, S. M. (1996). *Essential Psychopharmacology. Neuroscientific Basis and Practical Applications.* Cambridge University Press, Cambridge.

Stassen, H. H. *et al.* (1988). Familial syndrome patterns in schizophrenia, schizoaffective disorder, mania and depression. *European Archives of Psychiatry and Neurological Science*, **237**, 115–123.

Stoltenberg, S. F. and Burmeister, M. (2000). Recent progress in psychiatric genetics: Some hope but no hype. *Human Molecular Genetics*, **9**, 927–935.

Zuckerman, M. (1995). Good and bad humors: Biochemical basis of personality and its disorders. *Psychological Science*, **6**, 325–332.

Chapter 4: The storm-tossed soul

Deakin, J. F. (1996). 5-HT, antidepressant drugs and the psychosocial origins of depression. *Journal of Psychopharmacology*, **10**, 31–38.

Egeland, J. A. *et al.* (1983). Amish study I: Affective disorders amongst the Amish. *American Journal of Psychiatry*, **140**, 56–61.

Gilbert, P. (1992). *Depression: The Evolution of Powerlessness.* Erlbaum, Hove.

Jamison, K. R. (1993). *Touched with Fire: Manic-Depressive*

Illness and the Artistic Temperament. The Free Press, New York.

LeDoux, J. (1998). *The Emotional Brain.* Simon & Schuster, New York.

McGuire, M. and Troisi, A. (1998). *Darwinian Psychiatry.* Oxford University Press, New York.

Nemeroff, C. B. (1998). The neurobiology of depression. *Scientific American,* June, 28–35.

Ostwald, P. (1985). *Schumann: Music and Madness.* Victor Gollancz, London.

Raleigh, M. J. *et al.* (1984). Social and environmental influences on blood serotonin concentrations in monkeys. *Archives of General Psychiatry,* **41**, 405–410.

Raleigh, M. J. *et al.* (1991). Serotonergic mechanisms promote dominance acquisition in adult male vervet monkeys. *Brain Research,* **559**, 181–190.

Whybrow, P. C. (1997). *A Mood Apart: A Thinker's Guide to Emotion and its Disorders.* Basic Books, New York.

Chapter 5: The sleep of reason produces monsters

Abi-Dargham, A. *et al.* (2000). Increased baseline occupancy of D_2 receptors by dopamine in schizophrenia. *Proceedings of the National Academy of Sciences of the USA,* **97**, 8104–8109.

Andreasen, N. C (2000). Schizophrenia: The fundamental questions. *Brain Research Reviews,* **31**, 106–112.

Andreasen, N. C., Paradiso, S., and O'Leary, D. (1998). 'Cognitive Dysmetria' as an integrative theory of schizophrenia: A dysfunction in cortical–subcortical–cerebellar circuitry? *Schizophrenia Bulletin,* **24**, 203–218.

Bebbington, P. and McGuffin, P. (1988). *Schizophrenia: The Major Issues.* Heinemann, Oxford.

Bentall, R. P., Jackson, H. F., and Pilgrim, D. (1988). Abandoning the concept of 'schizophrenia': Some implications of validity arguments for psychological research into psychotic phenomena. *British Journal of Clinical Psychology*, **27**, 303–324.

Chua, S. E. and McKenna, P. J. (1995). Schizophrenia—A brain disease? A critical review of structural and functional cerebral abnormality in the disorder. *British Journal of Psychiatry*, **166**, 563–582.

Elton, T. C. *et al.* (2000). Decreased Serotonin 2A receptor densities in neuroleptic naïve patients with schizophrenia: A study using [^{18}F]Setoperone. *American Journal of Psychiatry*, **157**, 1016–1018.

Frith, C. D. (1979). Consciousness, information processing and schizophrenia. *British Journal of Psychiatry*, **134**, 225–235.

Frith, C. D. (1992). *The Cognitive Neuropsychology of Schizophrenia*. LEA, Hove.

McGrath, M. E. (1984). Where did I go? *Schizophrenia Bulletin*, **10**, 638–640.

Sass, L. A. (1992). *Madness and Modernism*. Basic Books, New York.

Spoerri, E. (ed.) (1997). *Adolf Wölfli: Draftsman, Writer, Poet, Composer*. Cornell University Press, Ithaca.

Tononi, G. and Edelmann, G. M. (2000). Schizophrenia and the mechanisms of conscious integration. *Brain Research Reviews*, **31**, 391–400.

Woody, E. and Claridge, G. (1977). Psychoticism and thinking. *British Journal of Social and Clinical Psychology*, **16**, 241–248.

Zakzanis, K. K. and Hansen, K. T. (1998). Dopamine D2 densities and the schizophrenic brain. *Schizophrenia Research*, **32**, 201–206.

Chapter 6: Such tricks hath strong imagination

Andreasen, N. C. (1987). Creativity and mental illness: Prevalence rates in writers and their first-degree relatives. *American Journal of Psychiatry*, **144**, 1288–1292.

Dykes, M. and McGhie, A. (1976). A comparative study of attentional strategies of schizophrenic and highly creative normal subjects. *British Journal of Psychiatry*, **128**, 50–56.

Heston, J. J. (1966). Psychiatric disorders in foster home reared children of schizophrenic mothers. *British Journal of Psychiatry*, **112**, 819–825.

Jamison, K. R. (1993). *Touched with Fire: Manic-Depressive Illness and the Artistic Temperament.* The Free Press, New York.

Karlsson, J. L. (1970). Genetic association of giftedness and creativity with schizophrenia. *Hereditas*, **66**, 177–181.

Karlsson, J. L. (1984). Creative intelligence in relatives of mental patients. *Hereditas*, **100**, 83–86.

Keefe, J. A. and Magano, P. A. (1980). Creativity and schizophrenia: An equivalence of cognitive processes. *Journal of Abnormal Psychology*, **89**, 390–398.

Ludwig, A. (1995). *The Price of Greatness: Resolving the Creativity and Madness Controversy.* Guildford Press, New York.

McNeil, T. (1971). Prebirth and postbirth influence on the relationship between creative ability and recorded mental illness. *Journal of Personality*, **39**, 391–406.

Post, F. (1994). Creativity and psychopathology: A study of 291 world-famous men. *British Journal of Psychiatry*, **165**, 22–34.

Richards, R., Kinney, D. K., and Lunde, I. (1988). Creativity in manic-depressives, cyclothymes, their normal relatives, and control subjects. *Journal of Abnormal Psychology*, **97**, 281–288.

Simonton, D. K. (1988). Age and outstanding achievement: what do we know after a century of research? *Psychological Bulletin*, **104**, 251–267.

Woody, E. and Claridge, G. (1977). Psychoticism and thinking. *British Journal of Social and Clinical Psychology*, **16**, 241–248.

Chapter 7: The lunatic, the lover, and the poet

Cronin, H. (1991). *The Ant and the Peacock. Altruism and Sexual Selection from Darwin to Today.* Cambridge University Press, Cambridge.

Finnegan, R. (1977). *Oral Poetry: Its Nature, Significance and Social Context.* Cambridge University Press, Cambridge.

Miller, G. (2000). *The Mating Mind.* Heinemann, Oxford.

Rasmussen, K. (1931). *The Netsilik Eskimos: Social Life and Spiritual Culture.* Report of the Fifth Thule Expedition, 1921–24, Vol. 8. Gyldendalske Boghandel, Copenhagen.

Roberts, H. H. and Jenness, D. (1925). *Songs of the Copper Eskimos.* Report of the Canadian Arctic Expedition, Vol. 14. Canadian Arctic Expedition, Ottawa.

Zahavi, A. and Zahavi, A. (1997). *The Handicap Principle: A Missing Piece of Darwin's Puzzle.* Oxford University Press, Oxford.

Chapter 8: Civilization and its discontents

Freud, S. (1930). *Civilization and its Discontents.* Hogarth Press, London.

Hare, E. (1983). Was insanity on the increase? *British Journal of Psychiatry*, **142**, 439–455.

James, O. (1998). *Britain on the Couch: Treating a Low Serotonin Society.* Random House, London.

Murphy, H. B. and Raman, A. C. (1971). The chronicity of

schizophrenia in indigenous tropical peoples. *British Journal of Psychiatry*, **118**, 489–497.

Sartorius, N. *et al.* (1978). Cross-cultural differences in the short-term prognosis of schizophrenic psychoses. *Schizophrenia Bulletin*, **4**, 102–113.

Sartorius, N. *et al.* (1986). Early manifestations and first-contact incidence of schizophrenia in different cultures. *Psychological Medicine*, **16**, 909–928.

Stevens, A. and Price, J. (1996). *Evolutionary Psychiatry: A New Beginning*. Routledge, London.

Torrey, E. F. (1987). Prevalence studies in schizophrenia. *British Journal of Psychiatry*, **150**, 598–608.

Chapter 9: Staying sane

Argyle, M. (1987). *The Psychology of Happiness*. Routledge, London.

Csikzentmihalyi, M. (1997). *Living Well: The Psychology of Everyday Life*. Basic Books, New York.

Jamison, K. R. (1993). *Touched with Fire: Manic-Depressive Illness and the Artistic Temperament*. The Free Press, New York.

Marshall, M. H. *et al.* (1970). Lithium, creativity and manic-depressive illness: Review and prospectus. *Psychosomatics*, **11**, 406–488.

Schou, M. (1979). Artistic productivity and lithium prophylaxis in manic-depressive illness. *British Journal of Psychiatry*, **135**, 97–103.

Thayer, R. E. (1996). *The Origin of Everyday Moods*. Oxford University Press, New York.

Index

Abi-Darghem, Anissa 128
adoption studies 55
affective disorder 20–3, 50–2, 95–100, 109–112
  bipolar 20, 22
  brain mechanisms 77–8, 98–9, 100, 133
  creativity and 140–5, 151–5
  genetic factors 50–2, 54–5, 60–5, 71–4
  neurotransmitters in 77–8, 98–9, 100
  outcome 22–3, 132
  prevalence 22, 52, 110–11, 194–5, 196–200
  relationship to schizophrenia 23, 85–9
  seasonal pattern 106–7
  symptoms 20–1
  treatment 45–6
  triggers 48, 100–1
  unipolar 20
alcoholism 20, 111
Alzheimer's disease 7
amfetamines 126, 208
anaemia, sickle cell 209
Andreasen, Nancy 143, 150
anhedonia 20, 111
anorexia 20, 111
anthropology 39, 183–6
antidepressants 22, 47, 77, 80, 107–8, 203
  discovery of 47
  mechanism of action of 47, 77, 80
'anti-psychiatry' movement 31, 43–5
antipsychotics 19, 46–7, 79, 118, 126
As I Lay Dying (Faulkner) 186
asylums 31–2, 188
attention, selective, in schizophrenia 123–6, 130–1

Baron, Miron 61
Barondes, Samuel 60
Bateson, Gregory 41–2
behaviourism 38–9
bereavement 25, 26, 27
bipolar disorder, *see* affective disorder
Bleuler, Eugene 19
Bourdieu, Pierre 174
brain
  abnormalities in psychosis 13–16, 32–3, 47, 77–8, 129–31
  basis of personality 83–5
  mechanisms of affective disorder 77–8, 98–9
  mechanisms of schizophrenia 126–31
  neurotransmission 74–81
  scanning 14, 127, 128, 131
Britain on the Couch (James) 193
bulimia 20, 111
Byron, George 10

Cade, John 45–6
Carlsson, Arvid 47
Carmen (Bizet) 178
Chapman, Jean and Loren 68
Chirico, Giorgio di 135–6
chlorpromazine 46
Civilization and its Discontents (Freud) 189
Claridge, Gordon 34, 64–5, 70
cocaine 126, 208
Coming of Age in Samoa (Mead) 39
consciousness 123–4, 130, 134
cortisol 98–9
creativity 141–51, 164–5, 167–8, 173–7, 210–12
  evolution of 173–7, 201
  and psychoticism 141–51
  and schizotypy 155–8

and thymotypy 151–5
Crow (Hughes) 214–15
culture 173–7, 178–80
  Inuit 161–7
  origins of 173–7, 180–1
cultural relativism 39, 40, 184–6
Cystic Fibrosis 58–60

D_2 (dopamine) receptor 127–8
D4DR gene 81–2
Dali, Salvador 137
Darwin, Charles 36, 45, 169
delusions 9, 19, 21, 27, 30–1, 119
dementia praecox 18, 116, 117; *see also* schizophrenia
depression 20–3, 96, 99, 109–12, 154–5, 194–5; *see also* affective disorder
  brain mechanisms 99
  genetic factors 51–6
  prevalence 22, 109–12, 194–5
  symptoms 20–1
  treatment 47
  triggers 48, 100–01
Descent of Man (Darwin) 36, 45, 169
disease, concept of 24–5, 29–30, 34–5
divergent thinking 31, 119–22, 134–6
Divided Self, The (Laing) 43
'double-bind' theory of schizophrenia 41–2
dopamine 29, 47, 76–7, 79, 98, 126–9
  D_2 receptor 127–8
  D4DR gene 81–2
  and personality 29
  in schizophrenia 79, 126–9
drugs
  of abuse 111, 126, 128–9, 208
  psychiatric 19, 22, 45–7, 107–8, 126, 203–4
Dryden, John 10
dualism 16

eating disorders 20, 111
Ebstein, Richard 81
Egeland, Janice 60

emotions, functions of 104–7
environment of evolutionary adaptedness (EEA) 159–60, 189
epidemiology of mental illness 37, 52–6, 73, 109–12, 157–8, 187–8, 194–5, 196–200
Eskimos, *see* Inuit
eugenics 37–8
evolution
  human 101–9, 140, 159–60, 181, 182–6
  of culture 159–60, 173–7, 181–2
  of mood system 101–9
  of psychosis 137–8, 140, 159–60
  sexual selection 170–3, 180–2
Examination at the Womb Door (Hughes) 214–15
extraversion 68
Eysenck, Hans 68, 85

family environment 39, 41–2, 44
Faulkner, William 186
Fisher, Sir Ronald 170
fluoxetine 29, 78, 107–8, 204
Fortes, Meyer 188
Freud, Sigmund 17, 40, 41, 189
Frith, Christpopher 124
Fromm-Reichmann, Freida 41

Galton, Francis 36–7
genes
  in Cystic Fibrosis 58–60
  D4DR 81–2
  in personality 67, 81–3
  in risk of mental illness 60–5, 71–4
  serotonin transporter 82–3
  in sickle cell anaemia 209
genome lag hypothesis 189–92
glutamate 128–9
Gottesman, Irving 53

hallucinations 19, 27, 30, 119
hallucinogens 128–9
Handel, G.F. 151
handicap principle 171–2
happiness 95–6, 204–5
Haslam, John 3,7

height 65–6
Heston, J.J. 149–50
hormones 98–9
hunter-gatherers 162–5
Hughes, Ted 214
hyperthymia 151–4
hypomania 151–4

Illustrations of Madness (Haslam) 3,7
imagination 1–3, 7–8, 182–6
imipramine 47
introversion 68
Inuit 161–7
iproniazid 47

James, Oliver 193
Jamison, Kay Redfield 50, 90, 142–3, 151, 208
joy 95–6

Karlsson, Jon Love 150
Keefe, J.A. 155
Kraepelin, Emil 16–17, 36, 37, 116
Kuti, Fela 173

Melville, Herman 12
Mendel, Gregor 58
mental illness
  causes of 48–51, 60–5, 71–4, 100–1
  concept of 24
  continua of 25–30, 64–5, 85–9, 110–11
  genetic factors in 50–6, 60–5, 71–4
  physical basis of 13–16, 32–3, 47–8, 98–9, 126–31
  types 12–13, 16–17, 18–23
  in writers and artists 141–7
  see also psychosis, affective disorder, schizophrenia
Midsummer Night's Dream (Shakespeare) 1–3, 156–7
Miller, Geoffrey 173, 175–8, 185
Minnesota Multiphasic Personality Inventory (MMPI) 156
monkeys 102–3, 107–8
monoamines 47,76–8, 98
mood 95–8, 101–9, 151–5; *see also* affective disorder
My Breath (Orpingalik) 161, 166–7

Laing, R.D. 43–5, 49–50
Larkin, Philip 36, 45
lithium 22, 45–6, 47, 118, 203, 208
Lombroso, Cesare 10–11
Lowell, Robert 152, 201
LSD 128–9
Ludwig, Arnold 143, 209

McGrath, M.E. 113, 135
McGuire, M. 107
McNeil, T. 151
madness, *see* psychosis
Magano, P.A. 155
Malinowski, Bronislaw 39
manic-depression, *see* affective disorder
mania 21, 97, 100, 111; *see also* affective disorder
Matthews, Mr. 3–8, 9, 116–17, 125
Maudsley, Henry 134–5, 187
Mead, Margaret 39

nature, vs. nurture 36–7, 48, 50–7
neuroleptics, *see* antipsychotics
neurons 75, 84–5
neurosis 13
neuroticism 68, 69, 82
neurotransmitters 15, 28–9, 47–8, 75–81, 107–8, 127–9, 131–2
Nigeria 27
Nineteen Eighty-Four (Orwell) 40
noradrenline, *see* norepinephrine
norepinephrine 47, 77–8, 98
nurture, vs. nature 38–9, 40, 41–3, 43–5, 48, 49–50

Origin of the Species (Darwin) 169
Origins of Mental Illness (Claridge) 34
Orpingalik 161–2, 163, 165–7
Orwell, George 40
outsider art 114

peacock's tail 169
personality theory 28–9, 67–71, 81–5, 156
personality disorders 13
PET scanning 14, 127, 128, 131
phenoziathines 46, 126
Picasso, Pablo 153
poets 142–3, 144, 145, 147
positron emission tomography (PET) 14, 127, 128, 131
Post, Felix 46
Prozac (fluoxetine) 29, 78, 107–8, 204
Prozac Nation (Wurtzel) 24–5
Price, John 189
Price of Greatness, The (Ludwig) 143
primates 102–3, 107–8
psychiatry 11, 16–18, 24, 31, 38–9, 44–5, 154–5, 203–4
psychoanalysis 40–1
psychometrics 28, 37, 67–71
psychosis 12–16, 18, 21–2, 34–5, 44–5, 48, 137, 202–3, 206
  boundaries of 12–13, 29–30, 34–5, 64–5, 85–9
  brain basis of 13–16, 32–3, 47–8, 77–8, 98–9, 126–31
  definition 12, 21–22
  functional 13–16
  genetic factors in 37, 50–7, 60–5, 71–4
  vs. neurosis 13
  organic 13, 16
  vs. psychoticism 71, 149, 207
  types 13, 16, 18, 85–9
psychoticism 68–71, 85–9, 139–40, 149, 207, 210–11

Raleigh, M. 107
Ramón y Cajal, Santiago 75
reserpine 46, 126
Richards, Ruth 148, 151
Roethke, Theodore 152–3
Rüdin, Ernst 37
runaway sexual selection 170–1
Rush, Benjamin 11
Ruskin, John 161

sadness 21, 25–6, 95
Sanity, Madness and the Family (Laing and Esterson) 44

Sass, Louis 121
schizoaffective disorder 23, 87
schizophrenia 18, 19–20, 30–3, 117–136, 204
  attentional deficits 123–6
  brain mechanisms 126–32
  causes, theories of 41–2, 43–5, 48–51, 52–6
  cognition in 119–22, 134–6
  creativity and 155–8
  dopamine in 79
  genetic factors 37, 52–6, 60–5, 71–4
  outcome 19, 132
  prevalence 20, 53, 187–9, 196–200
  relationship to affective disorder 23, 85–9
  symptoms 19, 119–20
  treatment 19, 46–7, 204
Schizophrenia Genesis (Gottesman) 53
'schizophrenogenic mother' 41
schizotypy 70, 85, 139, 155–8, 211; *see also* psychoticism
Schumann, Robert 90–5, 154
Seasonal Affective Disorder 106–7
Sedgewick, Peter 44
selection
  natural 138–9, 140
  sexual 170–3, 180–2
  stabilising 139, 140
serotonin 29, 46, 47, 77–8, 98, 99, 107–8, 111, 128–9, 194–5
  in affective disorder 77–8, 109, 110, 111, 194–5
  and mood 29, 98, 107–8
  in schizophrenia 79, 128–9
  transporter gene 82–3
sexual selection 170–3, 180–2
Shakespeare, William 1–3, 10, 156–7
shamanism 163–4
Simonton, D.K., 153
Skinner, B.F. 38
SSRIs, *see* Prozac
Stevens, Anthony 189
Szasz, Thomas 31–4, 117
suicide 23

Tennyson, Alfred Lord 50
Thomas, D.M. 40

thymotypy 85, 139, 151–5, 210–11;
  see also psychoticism
Touched with Fire (Jamison) 50
tricyclic antidepressants 47
twins 54–5, 56, 63–4, 149–50

unipolar disorder *see* affective
  disorder
Unquiet Mind, An (Jamison) 90

Van Gogh, Vincent 58

Velema (Fiji) 165
violent behaviour 20, 111

Walden Two (Skinner) 38
Watson, J.B. 38
White Hotel (Thomas) 40
Wölfli, Adolf 113–7, 135
writers 142–3, 144, 145, 147
Wurtzel, Elizabeth 24–5

Zahavi, Amotz 172